THE SAVAGE

D0462202

The Savage
and the Innocent

DAVID MAYBURY-LEWIS

Beacon Press   Boston

Beacon Press
25 Beacon Street
Boston, Massachusetts 02108

Beacon Press books
are published under the auspices of
the Unitarian Universalist Association of Congregations.

Second paperback edition published 1988 by Beacon Press
Printed in the United States of America

95 94 93 92 91 90 89 88   8 7 6 5 4 3 2 1

Library of Congress Cataloging-in-Publication Data

Maybury-Lewis, David.
    The savage and the innocent.
    "Second paperback edition"—T.p. verso.
    Includes index.
    1. Sherente Indians—Social life and customs.
2. Akwẽ-Shavante Indians—Social life and customs.
3. Indians of South America—Brazil—Social life and
customs.   I. Title.
F2520.1.S47M39 1988   981'.00498   87-42850
ISBN 0-8070-4603-5 (pbk.)

Contents

To our friends in Brazil

Illustrations

Preface to the Second Edition

This book describes our experiences among the Sherente and the Shavante Indians thirty years ago. At that time the Sherente lived in a remote backwater of Central Brazil. The Shavante hardly lived in Brazil at all, for the outside world had only just caught up with them—and not with all of them at that—and they still lived their nomadic lives according to their own lights and answerable only to themselves.

Since then Brazil has been transformed and its vast hinterland violently opened up in the name of development. The Brazilian model of development favored mining companies and large agribusinesses in the interior. It accentuated land hunger in a country larger than the continental United States (without Alaska) and worsened the conditions of the rural poor at the same time as it increased their numbers. It also put an end to the charmed isolation of the Indian peoples who had survived in the interior simply because they were remote and inaccessible enough to be left more or less alone.

The sleepy and ineffective Indian Protection Service described in this book was enlisted by authoritarian military regimes to see to it that the Indians did not stand in the way of development. As a result the agency was accused in the late sixties of collaborating to defraud and even annihilate the Indians it was supposed to be protecting. Brazil, startled to find itself accused internationally of practising genocide against its Indians, disbanded the Indian Protection Service and replaced it with FUNAI, the National Indian Foundation—but FUNAI was caught on the horns of the same dilemma. It was supposed to protect the rights of the Indians, yet the government expected it to find ways to advance "development" that was usually at the expense of Indian peoples.

The Sherente and the Shavante of course felt the effects of these momentous changes. We heard in the seventies of their battles to protect their lands and particularly of the fierce struggle that the Shavante had waged against the invading ranchers. They were no longer the naked warriors whom we had known. Warriors they still were, but now experienced enough to know that they could not hope simply to fight and win, and sophisticated enough to know how to back up their determination with politicking in the nation's capital.

The transfer of the capital from Rio de Janeiro to Brasilia took place at the time when we first went to work in Central Brazil. Brasilia was inaugurated while we were with the Shavante. In subsequent decades it accomplished what its founders had hoped—it moved Brazil's center of gravity inland by opening up the interior. Shavante country in the state of Mato Grosso was soon no longer beyond the effective national territory. Instead it became readily accessible to the new capital, a region that was making money by growing rice for the expanding population of Central Brazil. This put enormous pressure on the Shavante, for ranchers now wanted Indian lands, claiming that they could put them to productive use, instead of leaving them idle in the hands of savages.

The opening up of this part of Brazil also made it easier for the Shavante to get to Brasilia to argue their case, and this they did with increasing vehemence. Their truculence became legendary at a time when freedom of speech and of protest was sharply curtailed by the military for ordinary Brazilians. It was a Shavante going by the Brazilian name of Mario Juruna who caught the attention of Brazilian journalists by acquiring a tape recorder and recording the promises made to his people by government officials. That way he had proof when the Shavante were being lied to and did not hesitate to say so. This denunciation of official untruth delighted the press at a time when ordinary people heartily agreed that the government was lying to them, but did not dare to say so. When the first free elections for all offices excepting the president heralded the beginning of Brazil's return to democratic government, Mario Juruna was elected as a Federal Deputy from the state of Rio de Janeiro. So a man who had been a teenager chasing wild pigs beyond the frontier at the time when we first visited the Shavante became the first Indian ever elected to the Brazilian congress.

Meanwhile deputations of Shavante chiefs had become a common sight in Brasilia. They were helped by some FUNAI officials who were genuinely devoted to the Indian cause and they learned, as Brazil moved towards its political opening, that it helped them to take their case to opposition deputies and even to television networks that enjoyed giving coverage to Indians who stood up to the military. Eventually FUNAI was authorized by the government to demarcate and guarantee Shavante lands and to sponsor an elaborate rice-growing project among the Shavante themselves, so that the Indians could have a share, along with their neighbors, in the agricultural bonanza that had come to their part of the world.

It was these embattled but temporarily victorious Shavante that my wife, Pia, and I went back to in 1982. I had last visited them in 1967 and it was even longer since Pia had seen them. Our son Biorn, who had been with them as a toddler, was now grown up and married and we planned to bring him and his Japanese wife Hiroko with us as our assistants. But those plans went awry. On the eve of our departure from Brasilia Biorn was diagnosed as having cancer, so he and his wife returned to the United States for his treatment. Pia and I, not knowing what else to do, went forward into Shavante country, together with Bill Crawford, a professional photographer who wanted to make portraits of the Indians.

Two years later we revisited the Sherente. We had last seen them in 1963, when we had taken both of our sons, Biorn and Anthony, into the field with us. Biorn's cancer was now in remission, but neither he nor Anthony was free to accompany us on our return visit.

In the epilogue to this edition of *The Savage and the Innocent* I describe what it felt like to return after so many years to peoples we had known so well, and how they are faring as they strive to adapt their traditional ways to the modern world.

Preface to the First Edition

My wife and I lived among the Sherente for eight months in 1955–6 and among the Shavante for slightly longer in 1958. This book is an account of our experiences; it is not an essay in anthropology. Indeed I have tried to put down here many of those things which never get told in technical anthropological writings — our impressions of Central Brazil, our personal reactions to the various situations in which we found ourselves, and above all our feelings about the day-to-day business which is mysteriously known as 'doing fieldwork'. The narrative is therefore intentionally anecdotal. To those readers who find that this book is not as thrilling as a book about the wilds of Brazil should be, I offer my apologies. I can only add by way of explanation that every incident in it is true.

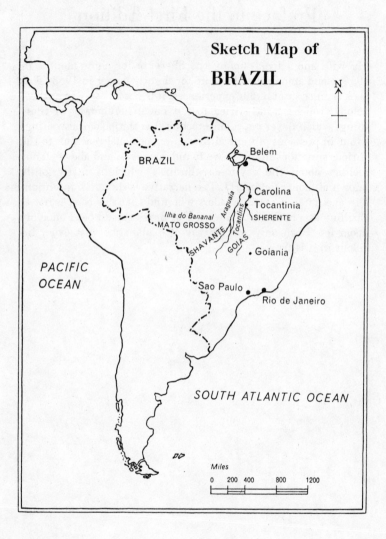

Sketch Map of
BRAZIL

N

BRAZIL

Belem

Carolina
Tocantinia
SHERENTE

Ilha do Bananal
MATO GROSSO

SHAVANTE

Araguaia

Tocantins

GOIAS

Goiania

PACIFIC
OCEAN

Sao Paulo

Rio de Janeiro

SOUTH ATLANTIC OCEAN

Miles

0   200 400   800   1200

Sherente

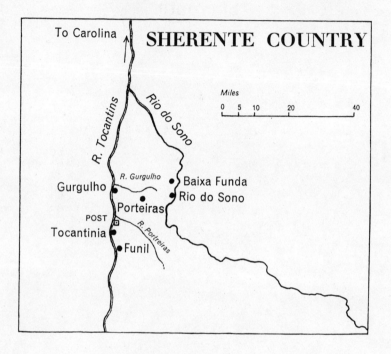

1

Method in our Madness

I am often asked why I went to Central Brazil, condescendingly by acquaintances, kindly by friends and in earnest by committees from whom I have requested research funds. Yet, like the foreigner who is asked 'How do you like our country?' every day by the well-meaning natives, I have found that repetition of the question does not make it any more answerable.

As an undergraduate I once took a course in the discovery, conquest and settlement of Spanish America. I marvelled then at the skill of the early transatlantic navigators and at the audacity of the conquistadors; but what intrigued me most were the first accounts of the American Indians. I conceived a romantic desire to know more about some of the people who had inspired such highly coloured narratives and who still, four hundred years later, seemed remote and exotic in a world jaded with travelogue.

It was with such highly unscientific ideas in mind that I wandered into an international conference, providentially being held at Cambridge, of anthropologists who specialized in American studies. With a self-confidence born of ignorance I went about buttonholing distinguished scholars and asking them if there were any way in which a young Englishman with a gift for languages and no money could go out to the Americas and do anthropological research. They listened to me, for the most part, with grave courtesy and offered me their good wishes.

Only one of them made a specific suggestion. He was a German professor of great personal charm who now holds a chair in Brazil. He assured me that there was a huge amount of work to be done in the interior of that country and a pathetic lack of people to do it. He would welcome me, he said, as his student, but he could offer me no advice as to how to finance my studies with him. 'In any case,' he concluded, 'if you want to mount an expedition, you might find it easier to go to British Guiana and start there. On that frontier nobody minds who goes in and out of Brazil. You could come and go among the tribes on the Brazilian

13

side of the border without troubling about all the bureaucratic delays and restrictions which would be imposed on you if you wanted to go into the interior from the south.'

I was more confused than ever. I had no funds and no very clear idea of what I was going to do beyond getting to know the Indians. Brazil was in any case the one South American country I had not seriously considered. After all, I spoke good Spanish and the language of Brazil was Portuguese. Yet I was already being advised to slip across a remote frontier to visit tribes I had never heard of as if this were the most natural thing in the world. Nevertheless the idea, with its piquant associations of the un-mapped and the unknown, appealed to me. Out of this slightly surrealist conversation came a succession of events which was to land me six years later together with my wife and a small baby in the midst of one of the most notoriously bellicose tribes of Mato Grosso.

When I went down from Cambridge I was still hunting for some means of supporting myself in the Americas while I in-dulged my new found interest in the Indians. I wrote letters to everybody I could think of, asking for a job in South America which would leave me enough time to get on with the study of anthropology, and I got accustomed to the flowery phrases of regret which took up most of the replies. Then, through a coincidence, I was offered a job teaching English in São Paulo, the very city where the professor of my acquaintance taught. It was exactly a year after our first conversation that I sought him out and reminded him that he had offered to teach me about Brazilian Indians.

I worked under him for the better part of two years, during which time I often wondered whether I would ever manage to get into the interior at all. Like so many Europeans my wife and I had imagined that Rio de Janeiro and São Paulo would be gate-ways to the enormous hinterland which fascinated us. We found instead that they faced outwards physically and metaphorically. São Paulo in particular was a huge, bustling metropolis deter-mined to out-buy, out-sell, out-build and outgrow every other city in South America before starting on the North American ones. It had tenuous and expensive communications with the more fertile parts of the interior. Its business men flew off to preside over enterprises mushrooming on the high plateaus of Central Brazil. Its cattle barons went to their uncountable herds in Mato Grosso in their private aeroplanes. The city tapped the resources of half a continent—and then did its best to forget it. The *sertanista*, or professional backwoodsman, a romantic figure with a beard and cartridges slung across him in bandoliers, could only be seen in the picture pages of the weekly magazines or, very occasionally, surrounded by elegant women at some cocktail party. The *sertanejo*, the man who lived his life in the backwoods, was a figure of fun who was never seen at all. To the popular mind he and the Indians of the interior represented all that was backward in Brazil—the Brazil from which everybody in the coastal cities, save romantics, scholars and sensation-seeking reporters, had resolutely turned away. It was, therefore, not much easier to mount an expedition from São Paulo than it would have been from Europe. If we were physically some thousands of miles nearer to our goal, we were psychologically much further removed from it. At least in Europe, or rather in England, wandering in the jungle is for some reason tolerated as a legitimate if not actually praiseworthy pursuit.

Nor was it so easy to leave after a couple of years in São Paulo. Our tiny flat had been our first home. We had made friends and had been more or less accepted by the neighbours, particularly by a delightful matron of uncertain age known to everybody as Dona Magdalena (even the telephone directory listed her without a surname) who had gone out of her way to be kind to us. She gave us ice at all hours during the panting hot months of January and February, and she used to invite us in to gargantuan meals whenever she thought we were not taking sufficient care of ourselves. Our friends telephoned her to leave messages for us, and she regularly gave them a detailed account of our movements, telling them what we were doing, with whom and why (often what we were likely to be eating as well). When her own friends came to visit her she would bang on our door and show us off in proprietary fashion so that they could see how a young European couple lived. This sometimes proved embarrassing as we had only the one room, and Dona Magdalena never minded what time she put us on display.

It seemed ludicrous and unprofessional to be filling up this minute flat with bizarre trinkets for Indians, about whom we had talked so much that they had come to seem more like figments of our imagination than people whom we were about to visit. There was a box full of knives: long stilettos of beautiful workmanship lying uneasily beside home-made hackers of the potato masher type. They had been seized by the police who made nightly rounds of the city trying to enforce the regulation that no one should carry a knife whose blade measured more than the width of three fingers, and someone in the museum had had the brilliant idea of asking for them so that they could be taken up by anthropologists as gifts for the Indians of the interior. There were combs and razor-blades (but not for shaving with), scissors and reels of nylon fishing line; there were hundreds of fishhooks and metres of cloth, together with mouth organs by the dozen and lipsticks for the men to paint themselves with. Dona Magdalena fluttered in and out of the flat as we collected our motley together, and we sometimes caught her praying for us when we went in to use her telephone. Meanwhile our friends plied us with gin tonics and hair-raising tales of what had happened to explorers they had once known.

We did our best to give an impression of calm and confidence, but it was somewhat spoiled by our ignorance of what was in store for us. We were going to work among the Sherente, about 1,200 miles away to the north, and all we knew about their present circumstances was that they were not *brabo* (wild Indians) but were on more or less peaceable terms with the local Brazilians. In this they reputedly differed sharply from the Shavante, erstwhile cousins of theirs, who still roamed the savannahs of Mato Grosso and fought off the encroaching settlers.

According to tradition and to the accounts of the early explorers who wandered down the Tocantins River into what is nowadays the state of Goiás, the Shavante and the Sherente had been a single people at the end of the eighteenth century. The most experienced travellers found it well-nigh impossible to distinguish between them. But they were all agreed on one point, namely that these Indians, be they Shavante or Sherente or Shavante–Sherente, were some of the most warlike and dangerous in the central uplands. They harried the gold and diamond miners who were the first settlers to come into this part of the country, and if troops were sent against them they replied by razing the mining settlements to the ground. Their nomadic style of life protected them from the full force of the reprisals which the settlers hysterically demanded, which was just as well, for some contemporary eyewitnesses were shocked by the brutality meted out to the Indians if the whites ever caught up with them.

At the end of the eighteenth century things appeared to take a turn for the better. An enlightened Portuguese governor made peaceful overtures to the Indians and offered to install them in villages under government administration if only they would cease their attacks on the settlers in his province. The result was that several thousand Shavante–Sherente marched on the capital of Goiás, scaring the inhabitants out of their wits and putting the governor in a quandary, for the little hamlet that was his capital could barely feed itself after a bad harvest, let alone entertain a whole tribe of visitors.

But the truce was a short-lived one. The governor hoped that, if he gave the Indians ready-built houses and a harvest to tide them over, they would settle down, till the soil and become good citizens. The Indians understood that they would live in the government's villages and be fed by the white men out of simple friendship. Many of them died in epidemics and more returned to the wilds when they realized that they would be obliged to work as labourers under government overseers. The whole scheme finally collapsed, for the governor's men embezzled the money set aside for the Indians and, instead of educating them, used them as slave labourers under military discipline.

Once more there was war between the Indians and the settlers, only this time it was embittered by the peace and by the Indian knowledge of the white man's ways. The Sherente, on the right bank of the Tocantins, went marauding as far as the State of Bahia, hundreds of miles to the east, but by the middle of the nineteenth century they were hemmed in by the settlers and forced to live in uneasy peace with their unwelcome white neighbours. The Shavante, on the other hand, remained intransigent. As more and more colonists poured into the mineral-bearing lands between the Tocantins and the Araguaia Rivers, they withdrew westwards and disappeared into the unmapped wilderness of Mato Grosso. They were not heard of again until the beginning of this century when frontiersmen discovered that if they lingered on the west side of the Araguaia they would sooner or later be attacked.

By the 1930s the expeditions of General Rondon had given a new impetus to the exploration of this part of Mato Grosso. The Shavante had been pushed back from the Araguaia, although they still appeared there occasionally, and were defending the lands to the west of the Rio das Mortes or River of Deaths. Journalists have made great play with this name, understandably so since the Shavante perpetrated a large number of killings along its banks, but its origin has nothing to do with the Indians.

Indeed, old atlases show that the river had already been renamed (it used to be called the gentle river, a reference to its slow current compared with the titanic Araguaia) in the days when the Shavante were still fighting the Brazilians hundreds of miles to the east in Goiás. It probably got its name from the days when gold was discovered in its shallows and a whole settlement of miners killed each other in internecine battle over the spoils. The people of Mato Grosso still believe that there is untold wealth to be found along its reaches, but probably the last outsider to take the legend seriously was Colonel Fawcett, who lost his life looking for Atlantis in the unpromising highlands.

We wanted to visit the Shavante, in spite of their awesome reputation, because they are one of a group of tribes which occupy the central uplands of Brazil and have done so for as long as anyone can remember. Some of these tribes had been studied, but only after their societies had been disrupted by the arrival of the white man with his twin gifts of land disputes and alcohol. Even so the accounts were fascinating. Here on the barren savannahs of Central Brazil these peoples, who used neither boats nor metals, who were ignorant of salt, tobacco and alcohol, who lived by gathering the roots and fruits of their native steppes and excelled only in war and on the hunt, had apparently developed social institutions of an astonishing complexity. The Sherente, for example, were reportedly divided into two groups or moieties. A person belonged to the group of his or her father and had to marry someone from the opposite moiety. All the men were further divided into two teams which ran races against each other on ceremonial occasions—races in which a two hundred pound length of tree-trunk was passed from shoulder to shoulder like a relay baton. The men also belonged to one or the other of a series of men's societies whose complicated interactions were the warp of tribal life. Further north the Eastern Timbira peoples seemed to have developed the moiety idea till their society sounded like variations around a single theme. They were said to have moieties which married each other and different moieties which performed important ceremonies. Their men's societies were ranged against each other in a moiety system and their male age-sets were organized into yet another system of moieties. The Western Timbira also had moieties, but these had nothing to do with marriage. Instead they were supposed to have four extra-ordinary marriage groups such that a man belonged to his father's group while a woman belonged to her mother's and membership in such a group obliged the individual to select a spouse from one other specified group.

This proliferation of moiety systems, associational and ceremonial institutions offered a golden opportunity for a comparative study. We hoped that, by visiting the Shavante, we would have an opportunity to see how one of these peoples operated their complex institutions and that this in turn would help us to understand the variations in the pattern, perhaps even to answer the crucial question as to why these tribes strove to organize their lives around a series of related dualisms. Our plan was to work first among the Sherente in the hope that, after a century of continual contact with the settlers, they would be able to speak Portuguese. This would help us to learn Sherente and, since Sherente was reputedly the same as Shavante, once we were reasonably fluent we could make a second visit to study the Shavante later.

But information about the Sherente was hard to come by. Everybody knew about the Shavante. They were news. The weeklies occasionally showed pictures of big naked men, a little out of focus, who appeared to be aiming bows or brandishing clubs at the intrepid photographer in his low-flying plane. These were copy from the latest visit to the River of Deaths. But the Sherente? Some said they had all died of smallpox. Others that the tribe had disintegrated and that its survivors were now indistinguishable from the local Brazilians. Some said they went naked and hunted with bows and arrows, others that they wore clothes and owned repeating rifles. The Indian Protection Service claimed that they were still numerous in the neighbourhood of Tocantinia, but that they clung obstinately to their Indian ways and were at loggerheads with the local settlers. The government urgently wanted somebody to go up there and bring back a report on conditions in the region. I volunteered for the task, for which there was no competition, and so it was that in June 1955 my wife and I travelled down to Rio where we were authorized to board a Brazilian Air Force plane for Carolina. From there we would have to make our own way up river to Sherente country and start work.

2

Neither here nor there

Rio de Janeiro has no winter. There is a season when the heat relents and the colours of the sea, the city and the tangled cliffs behind it are not cauterized in the hard white light of the sun. This is the dry season, when the nights are cool and the warm, cloudless days are free from impending rain.

It was in the last hour of such a June night that my wife, Pia, and I persuaded a sleepy taxi driver to take us out to the Galeão airport, ten miles away. We drove along the docks, each warehouse looming a little more distinctly as day took over from night. Soon we were in the outskirts of Rio where the early-morning traffic trickled along the edge of the bay like thin blood in an old man's veins. We passed a football field to landward where buzzards waited in long, silent rows. To seaward lay wisps of slums on the stagnant mud flats. It was hard to believe that people lived in these hutches, huddled beside a cement road, breathing the diesel fumes of the buses. Even harder to believe it a moment later when we pulled up before the main reception building on the airfield. Through long glass doors we caught a glimpse of smart boutiques and leather-covered armchairs. Under bold signs bearing the inscriptions 'Swissair—Zurich' or 'Pan American—New York' there were empty counters and idle weighing machines. The smart set did not travel at 5.00 a.m.

A cold ground mist eddied about our knees and wraiths of fog from the sea came swirling across the deserted runways. We were shown round the side to a concrete building with a dirty verandah and bare rooms without panes in the windows. Here, at last, there were signs of life. Two corporals of the Brazilian Air Force were on duty at a scarred table by a weighing machine that clearly did not belong to the same family as those in the reception building. They hoisted our baggage on to it and assured us with friendly indifference that we would have to leave half of it behind.

'Impossible,' I protested. 'We have permission to take 200 kilograms with us.'

I showed them our authorization to travel and the special baggage allowance we had been given. They were unimpressed.

'It is for the captain of the aircraft to decide,' we were told. No, it certainly is not, I thought; but it was clearly no use pursuing the subject then and there. I had yet to learn that the captain of a Brazilian Air Force plane is left to interpret his instructions and to juggle his ballast as best he can, once he is out on a mission. It was a system for which we had many occasions to be thankful when we were in the back of beyond.

On this occasion our priority was high. Not only did we get on to the plane, but the commander of the aircraft directed that all our baggage should accompany us. We settled down on the metal seats which ran the length of each side while the assortment of bundles and boxes which made up the passengers' luggage was lashed down, inches from our toes. A few moments later we were climbing steeply towards the mountains. The city, as you see it on the travel posters, all white buildings and glistening beaches, whirled beneath us in the first rays of the sun and then, suddenly, dipped out of sight behind the Serra do Mar, the Mountains of the Sea, which lapped upwards in wave after wave towards the high plateau of Central Brazil.

Draughts whistled through the plane and the metal seats were soon too cold to touch with comfort. We cursed our lack of foresight in having packed our blankets in our luggage, not knowing that one of the normal hazards of military air travel in Brazil is the impossibility of maintaining a comfortable temperature. The aircraft bakes on the ground, to the grave discomfort of its passengers, and then cools off much too thoroughly as soon as it is in the air again. But as we had not yet experienced the baking we were unappreciative of its antidote. It was a long time before we managed to fall into a sort of queasy sleep and we were at once woken up again as the plane began to lose height. It bumped down on to a scarlet runway and stopped in a cloud of dust before a solitary airport building.

We got out, and the heat struck us in the face. A moment later we were blinded by the darkness inside what passed for a terminal. It was deserted except for a solitary girl behind a counter, frying unappetizing looking sausages, all knotted and gnarled like lengths of shag tobacco. There were a few cane chairs scattered about the stone floor, and we made our way to them, while the aircrew gathered round the girl to see if they could persuade her to make some coffee. But she merely blushed and hung her head and went on stabbing away at the sausages as if they were flirtatious airmen.

Pia and I settled down in the creaking chairs and stretched our legs into the sun. We could feel the warmth coming into us like an electric current and with somewhat the same sensations of stimulation and unpleasantness. We had no idea where we were. We had nothing to do and there was nothing between us and the tiny blue hills on the horizon. Only a heat haze shimmering over the runway and a flat emptiness beyond it. We must have dozed for hours. Nobody was in any hurry. The passengers slept and the aircrew lounged in the shade. I awoke with a start and realized from the sun that it must be about midday. Still nobody moved. No plane had come and none had left. The girl was frying more sausages.

I went over for some coffee and took the opportunity to ask the captain of the plane when we would be leaving.

'I don't know,' he replied. 'One of our oil leads is leaking. I have radioed for a replacement and we shall just have to wait here until it comes.'

We waited all through the hot afternoon, while the girl plied us with sausages and coffee. Occasionally there was a flurry of activity. A jeep-load of passengers would come careering out of the distant town and wait for their plane. Usually it was from one of the commercial airlines. It would come in, ponderously slow in the clear air, and cover us with scarlet dust till we all looked like Martians escaped from a space film. Not one of them brought us an oil-lead. Their crews were smart in their white shirts, but, we thought, a trifle patronizing. The hostesses swayed their hips disdainfully at us. Then, with a flick of ailerons and another storm of dust, they would leave us, rather as a dog kicks rubbish over its own droppings. In the late afternoon our crew commandeered a station wagon and took us all downtown to make a night of it.

It was not even a one-horse town. Only donkeys idled in the unpaved streets, and a solitary bar was doing something less than a roaring trade. There were two boarding houses in the place, and they put us up reluctantly, as if we came bearing some contagion from far-off Rio. The town's amenities seemed to consist of a single barber's shop, where all the men had their hair cut. The women passengers, lacking such entertainment, were forced to spend the evening 'at home'.

The next day dawned even hotter than the last. We drank coffee. We bought fruit. We had our shoes cleaned. We could think of nothing else to do. The station wagon was supposed to pick us up again, but it was already an hour overdue. There was no sign of our aircrew. The Passenger with Initiative went to telephone for a vehicle. He was the sort of man who would have run the sports committee on an English ship, and I was afraid that he would organize team games among us if we were there much longer. He came back with his self-confidence only slightly impaired to tell us that he had only been able to find one telephone, in the doctor's consulting room. When he had made as if to go in and use it, the queue of patients had looked at him so threateningly that his courage failed him and he had fled. He lost much face with us, for a fat and friendly man who had done nothing until then save munch bananas and admire the shine on his shoes suddenly got us transport. He went across the road to a stationary bus which was about to set out on its route and, after a little haggling, persuaded the driver to come with him 'just for half an hour' and take us all out to the airfield at two shillings per head.

We collected our crew and returned to the girl with the sausages who told us that the relief plane had arrived unheralded in the early morning and gone again. We were dismally contemplating another day at the mercy of the Man with Initiative when a *deus ex machina* appeared.

One of the passengers, choosing his moment carefully, came forward with a dramatic announcement.

'I spoke to the captain of the other plane,' he said proudly. The stage was his and he knew it. He did not divulge his information all at once, but let it out bit by bit, enjoying each moment of the limelight. No, he had not gone into town with the rest of us. Wasn't it lucky that he hadn't? Besides, how could he afford to stay in a hotel, when the plane wasn't even supposed to stop? It was supposed to reach Belem the previous night, wasn't it? Well then . . . By rights he should have enjoyed a good night's rest in Belem, instead of sitting up in a chair at the airport. He had relatives in Belem and by now they must be very worried. The other plane? He was coming to that. It had not stayed for long. Not even stopped the motors. Just touched down and took off again. Yes, he had spoken to the crew, or rather they had spoken to him. They had asked him where the captain of the Air Force plane was, the one that was in difficulties. He told them that he was in town with the other passengers. Then they had given him some pieces of machinery that he did not understand and he had put them on the other side of the hangar in the shade, because it gets very hot during the day and he was not to know when the rest of us would be back . . .

The plane was finally two days overdue in Belem, but Pia and I reached our destination that same night. We entered Carolina, the 'third largest city in the State of Maranhão' as its inhabitants were proud to inform you, in the gathering dusk. We were all on the back of a lorry, except for Pia who was privileged to ride beside the driver, and, as we lurched over the rutted road, it was clear that we were in the north at last. The night was still glutinous with accumulated heat and fireflies danced beside the road. They seemed brighter than the houses which we were now passing, each one lit by a tiny kerosene candle showing a tangle of hammocks through the open door. They had no windows. We had to bend down as the vehicle passed under the boughs of huge trees growing in the middle of the road, and all the while we were coming nearer and nearer to a strange caterwauling which I gathered was at the centre of the town. It was not till we swung into the main street with its two bars and fitful electric light, together with the saloon noises that are the characteristic background for all towns in the Brazilian interior, that I realized what the central feature of the din was. The voice of a female crooner was blaring away from the town's loudspeaker. The song had been a hit in Rio the previous year and she was mouthing it with all the lethargic passion she could muster. The vagaries of the local generator served, however, to slow her down unintentionally at intervals during her rendition so that the total effect was grotesque beyond words. It was quite a relief when the announcer came on with advertisements and local gossip. It was not until I had had some experience of the eerie stillness of the bush that I came to understand why the town loudspeaker with its nightly hour (or two hours in affluent towns) of earsplitting noise was so important to the people who lived in the interior. It was their way of exorcising the jungle, of shouting defiance at the vast savannahs and claiming kinship with the cities.

A month later we were still in Carolina. In Rio there had been a last-minute contretemps about our permit to work among the Indians. The Indian Protection Service had not got around to issuing it, and the only man who could sign it was out of town. It had taken so much endeavour to get our flights to Carolina authorized that we could not bear the prospect of missing the plane and starting the whole procedure over again. So we accepted an assurance that the permit would be sent after us by the very next aircraft and took off for the wilds. Since Carolina is strategically located at a point where many Brazilian and even international air routes cross, there was no dearth of planes coming in with mail. After about a week we began to wonder whether they had forgotten about the permit. After a fortnight I wired to the coast. After three weeks we got a letter saying that it would be signed any day.

In the meantime we had become a feature of the local landscape and could look back with amusement at the inexperienced imaginings which had troubled our early days in the town. On our first night in the hotel we had been shown into a small, airless room whose sole furniture was a hard-looking double bed. It was lit by a naked bulb dangling from the ceiling, giving a pale splodge of light which left the walls and corners cobwebby with shadows. We could hear something rustling up where we judged the roof to be and before the night was very old we discovered that the mud walls did not reach to the ceiling, for every move and grunt from our neighbours on either side was clearly audible. Pia remembered stories of tarantulas which descended from the thatch and preyed on passing travellers, and I wondered where on earth I could put the funds for our expedition, all of which were lodged in my wallet. We flung open the boards which served us for shutters, exposing a hole in the wall which was to be our window. I have to admit that that night we slept with my revolver beside our heads, prepared for such eventualities as the travel books had led us to expect.

I was awakened before dawn by somebody climbing in at the window and I very nearly shot him; but in the nick of time he spread his wings and flapped into the courtyard, which was as well for us because it would have been an expensive business to shoot one of the landlord's *cocás*. These turkey-like birds were raised, together with innumerable chickens, right outside our windows. After that we learned to sleep with our windows closed so as to avoid being dazzled by the early sunlight and invaded by the local poultry. But we could not close our ears to the extraordinary garglings and hawkings which started each day for us. They came from the employees of the airlines and the telegraph company who were the only regular occupants of the hotel. These young men were stationed for tours of duty in Carolina, and when they were not at work they strolled around in their pyjamas, a practical notion which we could not copy since our expedition baggage contained no such frills. Their chief relaxations were playing pool and brushing their teeth, but the noise of their billiard balls did not concern us so much as their ablutions, which were performed with great thoroughness outside our door.

I think the Grand Hotel aspired to the name of Hotel largely on account of its having both pool table (which any bar might have) and rooms (which any pension might have). It was built round an open patio, where a few wilting plants were embedded in kerosene tins. A covered verandah ran all round the central space, being furnished on one side with tables to eat at and on two others with minute wash-basins whose waste, when it consented to run out at all, filtered sluggishly down a short pipe to be absorbed, but not too quickly, by the earth floor of the patio. On the fourth side were mud cubicles containing showers and latrines which could only be differentiated by their primitive fixtures, since both smelled abominably and were almost equally devoid of water.

After a few weeks in Carolina we came to welcome the presence of these showers as a luxury indeed. Every day, as we stood under their intermittent trickle, we would reflect that we had emerged more or less victorious from another afternoon of oppressive heat and enervating boredom.

In the long run it was the boredom which affected us most. It was aggravated by the tantalizing feeling that any day our papers might arrive and then we could really move into the backwoods and seek out the Sherente. This meant that we could not settle down to waiting. We had not yet learned the first lesson of the interior, which is to relax and make yourself comfortable wherever you are so that you may save your energies for a crisis. We plodded through the sandy streets till we had patrolled the town from end to end. We watched the motor-boats being unloaded down at the river and the women washing their clothes on the rocks. We watched the local craftsmen at work and had soft leather belts made for ourselves, a feature which, more than any other article of our equipment, was to mark us out as 'city folk' wherever we went. We hung around the post office when the mails came in and we hung around the airfield when the planes came in. We extracted a half promise that the tiny Air Force mail plane, which was the only one to go to Tocantinia in the heart of Sherente country, would give us a lift as soon as our permit arrived. We read all our books. But most of all we just talked.

At first these conversations followed the conventional pattern between strangers in the backwoods.

'Where are you from?' we would be asked.

'I am from England and my wife is Danish.'

This sometimes nonplussed our acquaintances, for in this part of Brazil foreigners are generally considered to be of two kinds— Americans and Germans. Most people had heard of the English, but Denmark was a distinct conversational hazard. Usually they would pass hastily to the next topic in the standard catechism.

'How many children have you?'

'We have not got any yet.'

'None at all? How long have you been married?'

'Two years.'

'Well, never mind. Don't give up hope. There will be plenty yet, you'll see.'

It was usually at this juncture that we were told an optimistic anecdote about someone who had not had any children, not even any born (a pathetic reference to the high infant mortality in these parts), and then after years of marriage had conceived at last.

After a few days we had been through this formal introduction with every notable in town. We had been classified, but henceforward it was hard to find topics of conversation. In shops or in private houses we were invariably asked to take a seat, and we would talk in a desultory way about Indians.

'Why do you want to go to the Indians? The Indian is an animal. He doesn't know what civilization and progress is. Do you know that until ten years ago the Canoeiro Indians used to attack travellers on this very river as they floated produce down to Carolina on their rafts? Yes, sir. The Canoeiros are really wild. They didn't leave this place alone till the aeroplanes started coming. Up till then you couldn't go into the forest without a gun and plenty of ammunition. That's why there was no progress in Carolina—none at all. It took two weeks to ride to Caxias and nearly a month to get to the capital of the state. Imagine! Nearly a month, and suffering all the way, what with Indians and the insects. Now nobody goes to São Luis any more. Why should we when Belem do Pará is only two hours away by plane? But down there in Goiás, that's where there isn't any progress. And do you know why? Well, I'll tell you. It's because of the Indians.'

Always a variation on the same theme. But I preferred it to the sycophantic interest which the town's scholars professed in the Indians. They were passionately interested, so they said, in the cranial measurements of the Indians, in the characteristics of their blood, in their religion or in whether they were descended from the lost tribes of Israel. They were, in fact, so interested in them that they wanted me to buy a whole host of things to take to them. And they were obviously relieved that Indians hardly ever put in an appearance in Carolina. Not that they would not be welcome if they did . . . but it was just more convenient this way, and it did not interfere with the progress of the town.

The town had indeed progressed so much that it had a club, which I was asked to address. I had no inkling of what was in store for me. I had merely been invited to a 'luncheon' by one of the dignitaries and had assumed that this was his pompous way of offering me hospitality. It was fortunate that Pia was wearing her one dress that day, otherwise we would have made an even worse impression than we did. I fear that the citizens of Carolina resented the fact that we tramped around their streets dressed up in slacks and boots as if we were expeditionaries—these were all the clothes we had—instead of wearing the linen suits and cotton dresses which they affected in order to differentiate themselves from the backwoodsmen. So I was doubly self-conscious when I discovered that I was guest of honour at this particular 'luncheon'. Pia was asked to hoist the Brazilian flag over the trestle tables, which she did amid a spatter of polite applause, while I could not help noticing that her tennis shoes were already a little worn and were definitely out of place. I was acutely conscious of the button which was missing from the top pocket of my bush shirt, of the array of linen suits around me and of the fact that I would clearly be asked to make a speech and could not think of a single thing to say. 'Carolina,' I thought frantically. 'I must say something about Carolina. What do I associate with Carolina? No . . . I can't say that. Of course. Progress.'

I rose to my feet still wondering how I was going to combine the topics of Indians, Carolina and progress. I said the usual things about how Europeans thought of Brazil as nothing but steaming jungles and often had very little idea of the big cities and exciting architecture to be found there. Even in São Paulo, I continued, warming to my theme, people did not know about the progress in places like Carolina. Who would have thought, for instance, that there would be a flourishing club here? I realized as I said it that the sentiment could be misinterpreted. I went on hastily to the subject of Indians, but I am afraid that the image got a bit blurred. My audience may well have got the impression that I had come to Carolina to study the Indians and that somehow Indians were the foundation of their civic progress.

It was not what politicians would call a 'successful' speech. We had the distinct feeling that there was a coolness towards us in the days that followed on the part of Carolina's business men. Only the smaller shopkeepers still called out *'Vamos sentar'* ('Come and sit down') when we strolled past their doorways. Luckily at this juncture our permit from the Indian Protection Service arrived.

All we had to do now was to wait for the mail plane, and then within an hour or two we would be among the Sherente—or so we thought. But we had reckoned without the wedding. It was the biggest thing that had ever been seen in that part of the state, and every bit of it took place within the precincts of the Grand Hotel. The religious ceremony was held in a chapel specially appointed in the siesta room. The civil ceremony took place on top of the pool table, got up in bunting for the occasion. Then there was an enormous spread on the verandah for everybody who was anybody in the town, and for those who were not, and for the women and children (who did not count) there was the wedding cake. This was a three-tiered affair which was much admired for a number of days and which I was instructed to photograph from a variety of angles. It was never consumed, and for all I know it may not even have been edible. But the crowning feature of the ceremonies was a supply of draught beer, a thing never known in Carolina before, which had been specially flown in from Rio by the comrades in arms of the bridegroom, who was a sergeant in the Air Force.

It was the beer that turned out to be our undoing. The next time the mail plane came through Carolina on its way north I talked to the captain and he assured me that he would take us and our kit down to Sherente country on the return journey, if he had room. I thought that the proviso was routine caution, but Pia was not so sure. She claimed to have detected a slight embarrassment in the aircrew and she was sure that the innkeeper was hiding something from us. We discovered what it was when the plane came back. The young lieutenant in command of it jumped out of the pilot's seat and disappeared into the radio room with suspicious haste. He had important business with the local detachment corporal. He had to scurry around, his revolver banging about on his thigh, seeing to the loading and the refuelling. It was clear that he was avoiding us.

I went up and reminded him that he had promised to take us with him. He was brusquely apologetic. He could not manage it this time. His load was too heavy and it was all top priority.

I was staggered. Surely he could squeeze us in somewhere, I insisted. After all, we were only going a couple of hundred miles to the south. By the time he came to the larger towns where he had important cargoes to take on, we would have been deposited at our destination.

The lieutenant was indignant. What did I expect him to do? He could not unload mails from his plane to make room for us. That was a very serious offence, for which he could be court-martialled. He knew we were scientists on a government mission and we could always count on the maximum of goodwill from him. But just now, with the best will in the world . . .

It was then that they started loading the empty beer barrels on to the plane. Pia had been expecting something of the sort and she was on to it in a flash. They had tried to put them in surreptitiously while I was engaged in conversation, but she came up and angrily pointed out what was happening. Fortunately her Portuguese failed her in her indignation. Meanwhile the lieutenant had lost face. He was angry with us for having caught him out in his little deception and he was itching to jump into his plane and put a few thousand feet between us. But he must have been nervous too. Certainly we were not very important, but we HAD got official papers from the government and we HAD caught him contravening Air Force regulations. Might we not be capable of making trouble for him? Reluctantly, he agreed to take us with him. But he was adamant about the baggage. Beer barrels or no beer barrels, he would not make room for it.

By now we knew enough about the ways of the interior to realize that there could be no question of 'having it sent on'. One of us would have to stay and bring it later. It only took us a few minutes to decide. I had to be the one to go ahead and present myself to the Indian Agent with the Sherente. Pia would have to come up river by boat. The lieutenant was already in his seat and was signalling that he was ready to go. Knowing that he would be only too pleased to go without me, I clambered in at the door with a parting injunction to Pia that she should try and stay with some American missionaries who had shown us some kindness in the past. A moment later we were in the air and I felt as ungallant as it was possible to feel. All the way to Tocantinia I crouched on the beer casks, sitting opposite the sergeant whose wedding had been the cause of the fracas. He found it necessary to study his log for the duration of the trip.

Five days later I was still waiting for Pia to catch up with me. She had promised to come on the first motor-boat which went up river. The trip would take her about four days, so we were expecting her any minute. I particularly wanted to be there when she arrived, so that she would not be on her own when she first saw the Indian Protection Service Post.

The post itself was invisible from the river which lay at this point like a great grey slab of water, bordered by miniature trees. It was only when a boat struggled in towards the bank that one could feel the racing currents under the surface of the Tocantins and could see the tall bluffs up which the river slithered every year to rest its saurian bulk among the tangled roots. From close inshore the trees took on their true perspective, and all the thicketry and snarled lianas seemed ready for a siege. On a bald patch of bank there stood a weather-beaten signboard as if connected by an invisible thread with the small boat which lay rotting at its moorings some thirty feet below. The boat rocked gently in the wavelets while the board teetered creakily in time with it. It was only when you scrambled up close that you could make out the inscription:

INDIAN PROTECTION SERVICE
8TH REGIONAL INSPECTORATE
TOCANTINIA INDIAN POST

A path led to a mud hut, half hidden by a huge tree.

When I arrived, there was a mulatto woman and her three tattered children waiting impassively at the top of the bluff. She was the agent's wife. Her husband, she said, was away but should be back by nightfall. She led me through the front room of their house, which was completely bare except for some harness on the floor, and offered me a seat at the table which was the entire furniture of the second room. I sat there all afternoon while she went quietly about her household chores and the children stood in a row and stared at me. At dusk the agent came in with a *mutum* he had killed, and I formally introduced myself and explained my business while we dined off the big bird.

Since then we had been waiting for Pia. Each day the agent showed me round the countryside and each night we sat swinging

in our hammocks while he told me about the Sherente. I had yet
to meet my first Sherente Indian. There were none anywhere near
the post. There had been, so the agent told me, but they found
there was no good land on which to plant near the agency and so
the whole village had moved fifteen miles inland. They were bad
Sherente at that village—idle, treacherous, always involved in
fights with their neighbours and with the settlers. Worst of all,
their chief, Pedro, claimed to be paramount chief of all the
Sherente and yet he was the worst Indian of them all. There
were others nearly as bad who lived to the south, about eighteen
miles away, the other side of Tocantinia, and there were some
who were rarely seen because they lived about fifty miles away to
the east, on the banks of the Rio do Sono (Sleepy River). Apparently, the only Sherente who were worth wasting time on
were those who lived to the north further down the Tocantins.
They were hard-working and sober. Their banana plantations
crowded down to the water's edge. Their manioc was plentiful,
their maize cobs fat and juicy. Their tobacco filled the slopes of the
river from the summer water-level right up to the wet-season
line. They were the ones I should visit first, he said. It did not
seem a bad idea at that.

Eduardo Pereira das Almas, to use the shortened form of his
full name, was only acting as the Indian Agent for the Sherente,
but he had been acting for a long time. The officials of the Protection Service are not exigent people, for they earn little and
usually they live in discomfort at places which are remote even by
Brazilian standards. Yet the agency for the Sherente was a post
which even they avoided like the plague. It was not on the fron-
tiers of settlement, so that it could have its own airstrip, but it
was far enough away to be conveniently forgotten. Indeed everybody would have been happy to banish the difficult and quarrelsome Sherente to some sort of limbo, where they could cease to
plague their neighbours and to worry the government. Since this
was not feasible, people did their best to forget about the Sherente
and the squabbles of the district of Tocantinia—except for the
local settlers who lived in a perpetual state of anti-Sherente indignation and the Indian agent who got no thanks from either
side. Eduardo incurred the enmity of all the influential people in
the miserable hamlet of Tocantinia and in return he drew a salary
of about four pounds a month. It was his favourite, one might
say his only, topic of conversation.

'It's not enough,' he was saying in his deep mulatto voice, 'it's not even enough for our food. I have to buy medicines for the children and rice when we have eaten our own. I have to buy coffee and sweetening. Then there's nothing left over, nothing at all and I still have to buy clothes and some ammunition for my gun.'

Why does he do it? I was thinking. The feeble light from our solitary wick dipped in kerosene accentuated the hollows under his eyes and the cavern of his mouth as he brought out again and again those syllables which were the leitmotif of his speech, '*Nǎo dá*. It's not enough.'

'Doesn't the government supply you with meat?' I asked.

'What a hope! There are a few steers here in the herd at the post, but I am not allowed to eat them. Only an occasional one. They are part of the assets of the place. I am supposed to raise cattle here. Raise cattle!' he went on bitterly, 'here where the grass is so mangy that they hardly get enough to eat! I can't get a cowboy to look after them on what the Protection Service pays, so they wander all over the place. Then the Indians kill them and eat them, saying they are theirs anyway, since they come from the Indian agency; the settlers complain that they wander in and spoil their crops and I cannot even get enough salt to keep them alive.'

'But surely you have salt here at the post?'

'Not a grain. Once a year the Service sends a lorry up from Goiania. If the rains come early, then it does not even reach us here—or if it breaks down on the way. If it gets here, then it brings salt and perhaps some harness for the horses and some petrol for the outboard motor. But the Sherente always know when the lorry is coming and they are all there to meet it. As far as they are concerned, it is their lorry which the government has sent for them. It makes no difference what we need here, I have to distribute something. If the government has sent some knives or tools, then I distribute them. If they haven't then I have to distribute some of the salt.'

'Then what do you do about the cattle,' I asked, 'if you have to give away the salt?'

'I try and buy salt on credit in Tocantinia; but nobody will give me much credit there. They know how much I earn and what the Service is good for. Still, nobody likes to see cattle die in these parts either. But very soon the shopkeepers are not going to care about the cattle or about me and my family either, for that matter. It's been two years now since we have had a lorry through and I have heard it said that they cannot get any driver to bring one up to Tocantinia. The last time there was a riot. The Indians attacked the lorry, trying to get at its contents, and the driver only narrowly escaped with his life. Nobody is going to drive all that way to be killed by a band of ungrateful Sherente.'

Eduardo stopped talking and listened intently for a moment. I could hear nothing but the clatter of banana fronds and the uneasy stirrings of his children in the next room.

'What is it?'

'Nothing. I thought I could hear a motor-boat, but it was nothing. There are not many boats that will travel by night in the dry season when the river is low. Besides Mundico told us there were no boats in Carolina when he left, so there can't be any coming for some days now.'

It was on the strength of this assurance that I went out the next morning with Eduardo in the hope of getting something for the pot. For hours we saw nothing at all, but we kept going because of our stomachs. They were rumbling suggestively as the sun rose towards its zenith, and the thought of a fat *mutum* for supper did something to offset the banal discomforts of the forest. We stopped to listen again for the call of the bird which we were following, or rather which Eduardo was following, for I don't think I could have tracked an elephant in those days, and it was then that we heard the noise. For a minute we stood stock still until there could be no further doubt at all. It was the regular mutter of a motor-boat approaching from the direction of Carolina.

'Quick, let's meet it by the river,' I said, and we set off, leaping and scrambling through the undergrowth, making detours and sometimes bullocking our way through dense thickets where there seemed no obvious way round. We were scratched and bleeding in no time. Eduardo was already wearing a tattered shirt and mine was soon in the same condition. The river was quite close by now, and the noise of the boat was unmistakably loud. It must have been level with us. We changed course and ran diagonally towards the water, hoping that we would come out on the bank before the boat had passed. But we were too late. As we emerged we could see the boat already fifty yards upstream and moving away from us. I thought I could make out Pia's figure, but I could not be sure. Anyway we shouted and waved our hats, but they could not hear us against the noise of the engines. I fired my revolver three times, but it was an old German officer's automatic weapon and it made a series of effete phuts which attracted no attention. The boat vanished round the bend of the stream and only the noise of its engines remained with us like a bad joke. We had missed our lift home. Worse still, Pia would be deposited at the foot of that desolate bluff with all our kit and she would find no one to meet her and only the staring family up in the mud hut to wile away the time with until our return.

We gave up the hunt for our *mutum* and walked disconsolately back as fast as we could go. It was late afternoon as we swung over the stile which kept the cattle away from Eduardo's house and tramped down the narrow path to the river. I could see the litter of our baggage where someone had dumped our things by the doorway. Pia was inside sitting on the bench. The children were staring at her. Her greeting was unprintable.

She had not had the best of trips. She had come on one of the smaller boats which had nothing but a platform at the back where the tiller was and where the cooking was done. In the covered body of the craft, all the available space was taken up by the cargo and the engines. The passengers had had to sling their hammocks over these. There was no room to sling them luxuriously in the proper way, so that one can lie diagonally across the voluminous cotton web and sleep in utmost comfort. Instead the passengers and crew lay like so many pilchards across the boat, hammocks touching and swaying in unison. Pia had been advised in Carolina that she should hobnob only with the captain of the boat, but he turned out to be a brusque and surly character who hobnobbed with nobody. However, a huge, good-natured negro had been friendly, had taught her the ways of the river and had slung his hammock next to hers so that they could pass the time in conversation. The first time the boat had stopped so that the men and women could split up and seek out separate bathing places along the river, there had been near panic at the women's pool when Pia appeared. Her hair was cut in the short style which was then the rage in São Paulo and she was wearing slacks. Since all women in the backwoods wear their hair and their skirts long, there was some momentary doubt as to whether she had come to the right place, and she was greeted with shouts of 'Women only!' Their food since they left Carolina had been rice and beans with little pieces of dried meat. She had, in fact, been looking forward to her arrival at the Indian Post which had been invested in her imagination with all the appointments of the ranches belonging to affluent Brazilians at which we had sometimes stayed in the south.

It was dark when she finished her story. We had been eating, together with Eduardo, and his amusement at Pia's impressions of river travel on the Tocantins did something to atone for the plain fare. There was no meat, thanks to our unsuccessful hunt. Eduardo's wife accepted our empty-handed return with the dark resignation of the backwoods. She went slowly back to her pots and did her best to cook some flavour into the mixture of manioc and rice which was to be our supper. She served us but did not join us. When the meal was over she heated some water and Eduardo brought it in to us in a shallow basin.

'Would you like to wash your feet?' he enquired. 'You sleep better if you do.'

It was true hospitality.

3

Sickness and sorcery

Three days later we reached the Sherente. A few came down to the river when our boat put in. Mostly women who stood about like toys without volition or expression. Railway lines of hair streamed bluely over their shoulders. A naked boy pranced about. They took a dispassionate interest in us. I was reminded of a crowd I had seen in Rio de Janeiro, gathered round a policeman, a pickpocket and the man whose pocket he had tried to pick. The young thief wept and struggled, begging to be given another chance. His pleading was obviously forced, an effort to get the crowd on his side, to play on the fathomless reserve of good humour which sometimes tempers tragedy in Brazil. The big, black policeman laughed and soothed him without for a moment relaxing the lock on his arm or permitting him to open his fist and rid himself of the damp little bundle of incriminating notes. The crowd waited, fascinated and determinedly aloof, empty-eyed as samba dancers at the height of carnival. So the Sherente received us—passively. They showed little interest as we unloaded our gear. I brought a ninety-pound sack of salt clumsily up the bank, trying to give an impression of muscular agility which I felt sure was belied by my performance. Still nobody moved.

Pia stood uncertainly among the women, and now there was a ripple of excited talk as they surged around her, fingering her clothes and her person, trying the texture of her shirt, her skin, her hair. Some men came sauntering down to see what was up. Their clothes, like those of the women, were the same as those of any backwoodsmen, but tattered and indescribably filthy. They wore their hair short. Only their markedly mongoloid features and the nasal bumble of their speech served to distinguish them from the settlers.

Eduardo went into action.

'Come on, you there, give a hand,' he exhorted. 'Can't you see that all these things are for you? The man has come from Rio, from the government.'

'The man' referred to me. The woman, after the fashion of the interior, was not referred to at all. Pia and I never did get used to the manner in which we were received all through Central Brazil. The senior male present would always step forward, shake me by the hand and pat me on the back in a formal embrace, saying at the same time '*Adeus, compadre, como vai?*' The first time I was greeted in this way I was considerably startled, for I translated the words literally as 'Goodbye, godson, how are you?' Later I realized that in these parts *adeus* still retained its old-fashioned meaning of 'God be with you' and that *compadre* was a conventional term of address used to bring strangers into the family circle, so to speak. After such an exchange we were usually led into the house, Pia being expected to follow where I led. Despite the unvarying repetition of this little ritual, habit often proved too strong for us and I would absentmindedly stand aside to let Pia go first. This placed our host in the awkward position of somehow circumambulating her and advancing to greet me without seeming discourteous.

On this occasion we had no precedent to guide us. These people looked like backwoodsmen, but they did not speak or act like them. On the other hand, they did not seem to act like Indians either, and anyway we were not quite sure how Indians *did* act in such circumstances, beyond realizing that each tribe (or as the Brazilians more aptly call them, each nation) would have its own greeting customs. We had not the knowledge to do the right thing confidently; yet, as would-be anthropologists, we were anxious not to do the wrong thing like insouciant explorers. So we hung back, isolated in our ignorance.

The magic words 'Rio' and 'government' finally produced a reaction. Young men shouldered our kit eagerly while the women crowded alongside.

'Is it true?' they asked. 'From Rio?'

'From Rio,' said Eduardo importantly. 'Come to learn all about the Sherente. That's why I brought them here first—to meet some good Sherente.'

'That's right,' came a chorus of assent. 'We are good Sherente. Here we work and we don't fight. The others only want to fight and live in the jungle like animals. We are not lazy. We don't get drunk. This is a village of plenty.'

It sounded like a litany and a smug one at that. I found out later that that was exactly what it was. Meanwhile we were walking through evidence of the Plenty which had been so often mentioned. The white path we were following was a crooked parting on the bald, brown scalp of earth. On both sides the ground was strewn with a riotous abandon of half-felled trees and rotting stumps. Clearly somebody or something had been active here. Perhaps an elephant's wrestling match? But there are no elephants in Brazil, I reminded myself, and realized at the same moment that Eduardo was talking.

'You see the amount of land they have cleared, *senhor*? These aren't Indians to live by begging. No, sir. All this will soon be planted and then in the spring you will see such crops! Now the whole place is full of bananas,' and he gestured vaguely over towards where I supposed the banana trees were.

'It's very impressive,' I said, playing for time. 'I'll mention it to the Indian Protection Service.'

I had clearly said the right thing.

'Yes, you do that.' Eduardo's eyes opened wide and his lower lip protruded as it did when he was particularly animated. 'You tell them that here on the Gurgulho (Gurgle) River the Sherente are really progressive.'

I shuddered at his choice of words.

There were about ten mud houses, eight of them drawn up in two lines facing each other across a plaza of trodden earth. At one end of this oblong stood the chief's house and the house of assembly, elegantly flanked by a row of palm trees. The other end would have opened on to park-like savannah had it not been partially screened by a spreading carob tree. The arrangement was aesthetically pleasing and academically exciting, for the Sherente had built their villages in sweeping semicircles in aboriginal times. The northern arc contained the houses of one moiety, which faced the houses of the other moiety located on the south side. From east to west the village was bisected by the 'path of the sun' as they called it.

Jacinto, the chief, was waiting to receive us. He was a big, shambling man with unruly grey hair which, together with his long face and nervous glance, gave him a curiously donnish air. He wore a straggly moustache as a sign of enlightenment. It served also to show that he was racially as well as culturally closer to the Brazilians than many of his fellow Indians, for 'pure' Amerindians, like orientals, have little facial or bodily hair. He embraced Eduardo after the fashion of the interior and then shook hands with us in the manner of a diplomat at a cocktail party, his eyes searching over our shoulders for something more profitable. He ushered us into the assembly room of the village and then went on to the guest of honour—our baggage. This had been dumped in the middle of the floor, and the villagers were already beginning to unpack it. We hastily made it clear that two of the bags contained our personal property and these were reluctantly surrendered for us to take into the back room where guests were lodged. It was a long, thatched airless cubbyhole. There was no light in it except when the door to it was left open, and what ventilation it got came through the cracks in the adobe walls and the holes in the roof. Here we slung our hammocks and grounded our bags.

When we emerged from our retreat a few minutes later the distribution of our largesse was in full swing. An old Caliban of a man, dressed in the remains of a sack, was issuing dishes of salt and cakes of brown sugar to the women and girls who crowded up to him. The men were perched like vultures along a pole which ran the length of the room and served as a sort of bench. They kept a sharp eye on those of their number who were sorting out the knives and the cloth, the fishing tackle and the bottles of cheap perfume. Jacinto sat by himself, disinterested and non-participant. He could afford to be. We had made him a special present in his capacity as chief.

We went over and joined him. At the same moment Eduardo came in with a sheaf of bananas—like a tourist guide, I thought, getting his cut for delivering the clients to a rapacious night club. He told us that Jacinto had agreed to house and feed us for the equivalent of about £2 10s. per month. This was a lot of money to the Sherente and a lot of money to most of the neighbouring Brazilians. Indeed, it was a lot of money to us, for our entire funds consisted of £160 lodged in my belt, and I was already beginning to tot up the cost of metres of cloth and steers for slaughter and to wonder whether we would be able to last out. Jacinto got the money at once, and his wife, in an access of all too audible glee, slaughtered a chicken for our supper. That is, it would have been for our supper, only most of Jacinto's family dropped in around midday to share it with us, so by the evening there was not much left. Not that we minded, for by that time we were too tired to eat. Eduardo, after the usual anxious minutes with the outboard motor went jammering away upstream in mid afternoon, and we spent the hottest part of the day ill supported by the sitting pole while Sherente clamoured about us. The men could speak a little Portuguese, just enough to cross-question us about ourselves, but not enough to understand the replies with any degree of certainty. Which was not surprising, for what reasonable account could we give for our presence among them? The women spoke little or no Portuguese, but could make their wants quite clearly understood without it. They wanted clothes. The bolder ones among them made it clear that they would have no objection if Pia stripped then and there and added what she was wearing to the kitty. Most of them just hung around in the doorway, blocking out the light, so that we sat in brown darkness and tried to classify the faces all around us. But they were nouns in a visual language to which we had no key. They came and went like the sounds of Sherente, whining and droning about us till our heads ached and we longed to be able to get away and pigeonhole these impressions in our minds.

I cannot recall how that day ended but I remember that we were awakened at some indeterminate hour of the night. Somebody was shouting, bawling some moonstruck phrase over and over. Then his hoarse voice unclotted and rushes of words came spilling into the darkness. He stopped, and the silence was so sudden and complete that we could hear the creaking of our hammock ropes responding to our breathing. The voice coughed, retched, gathered strength and was off again, gibbering into the night.

'What do you think it is?' Pia whispered.

'I don't know.'

'Do you think he's mad?'

The thought was in both our minds. Clearly I ought to go and find out, but I was swathed in my hammock. It clung to me, adapting itself to my body and holding me in that little womb of a room. I discovered that I was tired and that I had no wish to go out and stalk a lunatic. But I was here to learn about the Sherente, wasn't I? My conscience won and I struggled clumsily on to the floor. I could not find my tennis shoes or my torch. My rucksack felt unfamiliar. I cursed my own unpreparedness. The voice was fainter now. Perhaps the man was delirious?

It was chilly in the plaza compared with the stuffy hut. The darkness was muddied by approaching dawn. I could hear coughing from many of the huts, and from the far end of the village came the voice, hoarser than ever, burbling now like a protesting camel. I walked in that general direction, stubbing my toes on the roots and bumps in the unfamiliar ground. Still I could not locate the source of the noise and suddenly I was seized with embarrassment. What if somebody emerged from his hut and found me prowling around the village in the early morning? They might not take kindly to being spied on by their guests. And what if I did discover which hut the noise was coming from? I could hardly just barge in and announce that I had come to see what was going on. In any case it did not seem to emanate from any one house. I was certain it was coming from the second last one in the village on the left and set a course for it as best I could, but when I arrived there the voice was over to my left again. I had somehow passed it. The more I wandered after it, the more it eluded me, till I tired of this willow-the-wisp game, squatted down by the wall of one of the huts and waited.

It was getting light enough to make out vague shapes, and I was bitterly cold. But I had not long to wait. The shouter came stumping round the corner of the hut where I was sitting and passed within a couple of feet of me, ignoring me completely. It was Caliban who had presided over the distribution the previous day. He was naked except for a loin cloth, and he carried a club which he used as a walking stick or to pound on the ground while he spoke. He stopped close by where I could just discern his jerky movements and bellowed hoarsely into the thin air, emphasizing each burst of speech with a thud of his club. I realized that I was watching one of the elders haranguing the village. It crossed my mind that his speech might have something to do with our arrival, but I never found out if it had. The villagers assured me that he was admonishing them to respect their ancient customs and to obey their elders, but they were vague as to whether he said anything more specific or more topical. In any case they paid little attention to his exhortations, for the Sherente on the Gurgulho were determined to be progressive.

On that first morning Jacinto showed me round the village lands with seigneurial pride. We picked our way through the gutted chaos of the clearings and came out on to the long tobacco patches sloping down to the river.

'Why do you grow tobacco instead of food?' I asked innocently, remembering the previous day's complaints of how hungry the villagers were before the harvest.

Jacinto darted a quick look at me to see whether I was being sarcastic.

'We Sherente like to smoke,' he replied with a shy giggle. 'The civilisados sell us tobacco, but it is dear.' He drew out the word for emphasis. 'And when the Indian has anything to sell to the civilisado, then they don't want to give us anything for it.'

He had adopted the whining voice in which the Sherente intoned their wrongs, so I hastily changed the subject.

'Where are the bananas which I have heard so much about?'

We walked along the river under the rim of the bank, passed a lone Indian hoeing a row of maize and clambered up into the plantation. From the hot, sheltered calm of the river gulley we emerged into an embattled forest of palms. Their fronds rattled and chattered above the sprays of greenery which swarmed about their slender trunks. There were slitherings and scutterings in the undergrowth as we approached.

'Watch out for snakes,' said Jacinto amiably.

My feet suddenly felt exposed in their thin tennis shoes and I wished that I had not been too proud to put on my boots that morning. I let Jacinto go ahead in his home-made sandals. He swung his machete in an easy rhythmic stroke, and a young palm spilled bananas into my hands. Later we collected a sheaf of sugar canes and returned to the village along the thickly forested stream from which it took its name.

Pia was nowhere to be seen. The sun had almost reached that point where it hangs mesmerized, unmoving and unblinking over the numb earth. We shuffled gratefully into the darkened assembly room where some of the village dignitaries had already settled themselves on the pole preparatory to whiling away the hot hours of the day in conversation.

Sizapi, the village seer, was waiting for us together with Jacinto's brother, who was known as Dbakro, meaning literally 'wanton monkey'. His father must have had a sense of humour to have selected such a name for him, for he had the disgruntled expression of a mischievous anthropoid, saved from ugliness only by his fine dark eyes. He had been chief of the village before Jacinto, he told me, but he had abdicated because the position was not worth it. He never got a thing, neither payment nor prestige, only headaches, for the Sherente were so quarrelsome. So he gave it up and let Jacinto take over.

Jacinto shuffled his feet like a guilty witness and admitted that this was so.

'But I don't get anything out of it either,' he hastened to add. 'Where is my salary? It never comes. I never saw any uniform. I never saw any sword. In the old days, in the time of the great chief Sliemtoi who went to Rio to see the Emperor, a Sherente was proud to be a chief. He was paid a salary and he had a cocked hat and everybody respected him. Now, nobody listens to their chief any more.'

'But the Emperor is dead,' I said. 'There is no Emperor now, only the government in Rio, and it is they who have created the Indian Protection Service especially to look after the Indians.'

Dbakro spat demonstratively. Sizapi spat in imitation.

'The Service is no good,' they said, bringing the words out like voided phlegm. 'The Service does nothing for the Indian. It gives us nothing. When the government sends something for the Sherente it is the agents of the Service who steal it all before it gets to us. *They* make lots of money.'

I thought of Eduardo and his pitiable salary, but I let the remark pass. It was suddenly brought home to me that these tatterdemalion peons suffered from something closely approaching delusions of grandeur.

'We want our lands,' they kept repeating, and I jumped eagerly into the conversation, explaining that one of the tasks entrusted to me by the government was that of mapping the whole munici- pality of Tocantinia together with its Sherente settlements, so that part of it could be legally set aside as Indian land. But the old men were not listening.

'You are a good man,' they said, unconvincingly. 'You under- stand us. You will see that our lands are given back to us. We used to hunt over all the land between the Tocantins and the Rio do Sono. We had villages on the far bank of the Tocantins two days' journey downstream. Now there are no villages. The *civilisados* moved in and the Sherente had to go. When we fought they sent soldiers, and we fought them too. Then they made a treaty. Yes, sir! The Emperor himself gave us our own lands on a piece of paper. But the *civilisados* sent their cattle in to graze there. When we killed the cattle, they said we were thieves and that it was impossible to live near us. Now our villages are destroyed and the *civilisados* work unceasingly so that we shall be moved even from Tocantinia. You must tell them all this in Rio. You must tell the government that we want back the lands we used to hunt over, right up as far as Iron Pan.'

'But that's twenty leagues away,' I expostulated. 'Three days from your nearest village. There are towns between here and there. You are only a few Sherente now. What would you do with all that land?'

'It is ours,' they chanted. 'The Emperor said so. We want it back.'

I was relieved when Pia rescued me from this impasse. She came in with her hair plastered down and told me that she had been for a swim with the women. Jacinto got up.

'Let's go for a bathe,' he said. His smallest sons had been waiting for these words. They skipped out into the sun with their gourd water-bottles. Jacinto picked up the earthenware pitcher and followed them. I brought up the rear.

At this time of year the Gurgulho lived up to its name. It ran gently between steep banks, disassociating itself from the refuse of the rainy season which formed a jagged rim of compost higher up the slope. We entered it at the ford, where the tawny sand was ribbed beneath our feet, and let the water carry us out to the white stones of a natural pool. There we lolled in the shallows while the children dived and splashed around us. They were almond eyed and chocolate coloured, delicately beautiful through a happy combination of traits inherited from their Indian father and his second wife, a Brazilian. The smallest one tumbled about beneath me and suddenly leaped high out of the water like a startled fish, interrupting my reverie, in which I was wondering whether bathing was the only really sensual pleasure which Central Brazil had to offer. With the possible exception of sex? Among the neighbouring Brazilians sex was something furtive, quick and uncomfortable. Men talked about it and fought about it. Women moaned about it. One got the impression that people took their love as they took their food, a mess on a tin plate to be stirred up and gobbled. Perhaps among the Sherente it would be different. They might after all be the gentle, wronged people they gave themselves out to be. Nobody had ever tried to find out what sort of people they really were. Historians, travellers, settlers— all of them had constructed caricatures of the Sherente, cartoons based on the few salient traits which they, their enemies, had been quick to notice or to take for granted.

That night it was hard to believe any of them. The sun released us and disappeared in a blaze of improbable colour. There was an early moon which ennobled the huts and improved the whole village by making it slightly out of focus. Only the palm trees, swaying rhythmically in the new light, seemed to have come into their own. Jacinto sat on a cowhide stool outside his hut. His two wives, one Sherente and one Brazilian, were stretched on sleeping mats beside him, their children playing over them. Pia and I sat on stools and wished that we had sleeping mats instead. One by one the men of the village shuffled over to our little group.

Some of them were accompanied by their wives. Nearly all of them brought along a child to dandle. Their talk ebbed and flowed, unhurried and unforced. Much of it was in Portuguese for our benefit, though occasionally the speaker would switch to Sherente when he wished to emphasize some point. In Sherente they spoke sharply, breathlessly, the voice coming down like a conductor's baton at the end of each idea. Their Portuguese was weedling, but fluent and often surprisingly well phrased. They were good talkers and good speakers. I could sense the assurance and the rotundity of Dbakro's periods as his deep voice addressed the assembled company. Jacinto, surprisingly, said little except to interpret for us. His senior wife seemed to take more part in the conversation than he did.

'We shall tell you all about the old days,' Jacinto said to me in an aside. 'The old men here are the ones who will teach you. I am young and ignorant. I know nothing.' (This was a conventional self-deprecation. It was also true.)

'Tomorrow Wakuke will come to see you. He has studied much and he will tell you the truth about the Sherente. We will teach you our language and then you will be able to tell them about us in Rio . . . the truth, not the gossip which they hear every day by telegram from our enemies.'

He turned to one of the patches of shadow where two young men appeared to be lying on top of an undifferentiated bundle of women and children.

'Uncle,' he said. The answering grunt was clearly Caliban. 'Uncle, tell Wakuke about our factions.' He looked at me to see if the remark had registered. 'The people have decided to call you Wakuke,' he announced. I was flattered to be thus called after my guru, but for the moment I was too engrossed with Caliban's explanations to express my appreciation.

The old man spoke bad Portuguese. He coughed his retching cough and said shortly: 'We had factions, parties just like the *civilisados* have. They have parties which are always fighting. So, in the old days, the Sherente were divided into *kuze* and *krenprehi*, into *krozake* and *wairi*. They were always fighting, just like the *civilisados*. But that is all finished now.'

Sizapi, who prided himself on his knowledge of the outside world, took up the old man's tale.

'There were good parties, just like your democratic party, and there were bad parties like your Kemunis.' It took me a moment or two to realize that he meant communist.

One of the young men was bored with the recital. He stood up, stretched and dug his companion with the tip of his toe.

'*Are kto* (Let's go),' he said.

The heaps of figures writhed and emitted sleepy noises. Two or three men got up and strolled off into the darkness.

'I am going,' said the bored one in formal farewell.

'Go!' came the replies, like a salvo of pebbles.

Sizapi sighed. 'The young men are not interested any more. They don't know and they don't want to know.'

'Wakuke! *Are kto!* We are going hunting.'

Every morning now I would be awakened by this cry. The dry season was dragging to its end and there was little work to be done in the plantations. There was little food in the village either. Plenty is a relative matter in Central Brazil. So every day the men of the Gurgulho combed the surrounding country for meat.

I just had time to pull on my boots, take a mouthful of the horrible nut mash that Jacinto's family were eating, and join Kumseran on the trail down to the river. Women sat in their doorways with one leg folded under them, each one holding an axe blade upwards between their thighs. With idle ease they split the hard, dry babassu nuts on the blades and masticated the sapless kernels. They laughed at my nervous haste as they saw me hurry by.

'Where to?' they called, partly because it was customary and partly because everyone's business is public business in Indian villages. A failure to answer such a query, even when simply going into the jungle to relieve oneself, is construed as either sinister or surly.

'Hunting,' I replied, rather shortly for they already knew that very well. The answering laughter was sufficient indication of their views on my hunting capabilities. They laughed too because I was in such a hurry to catch up with the others. A Sherente would not have bothered. He would simply have followed their tracks at his own pace until he finally caught up with them while they were taking a rest or rifling some fruit tree. But I could not tell the difference between today's tracks and yesterday's, or last year's for that matter, over the hard ground. I had to be led like a child. The women knew it, and their gales of laughter as I sloshed across the Gurgulho at our bathing place made me even more bad tempered. It would be another long, hot, hungry day

47

and as likely as not we would come back empty-handed. We usually did.

Kumseran was waiting for me on the other side. He stood there patiently while I pulled on the clumsy boots which not so long ago had been my pride. I hated them now, for they forced me to go through this undressing and dressing process every time we crossed a stream and therefore to rely even more on the goodwill of the occasional Sherente who would wait for me while I did it. On the other hand, I did not dare to go barefoot or in tennis shoes. My feet were not hard enough and the distances we covered too long. One day I had hunted in plimsolls, sloshing through the streams like the rest of them and allowing my clothes and my footwear to dry on me as I walked. We must have covered about twenty-five miles that day and the following night I had such cramp in my feet that I could hardly sleep. So I had reverted to boots.

My heart warmed to Kumseran for his consideration. But he was known for his kindness and had some reputation as a peace-maker. This morning he was telling me about his child who was still running a high fever. I knew about it, for Pia had been with the little girl much of the night, but there was little we could do save fight off the crisis with codeine and hope that it did not kill her when it came.

'She saw the *romsiwamnarin*,' Kumseran was saying. 'She saw them coming to get her. She sat up and cried out and she was very, very hot.'

The *romsiwamnarin* are monsters who occur in most Sherente stories. Particularly, they beset travellers on the approaches to the village of the dead. So it was not surprising that Kumseran had been moved and upset by his daughter's vision. He had got up off his sleeping mat, drawn his three-foot machete and done battle with the monster then and there. The little girl, comforted by her father's protection, had relapsed into that horrible gasping sleep which exhausts the fever patient. Kumseran would eat no meat that day, for such strong food would harm his child in her present state. She might even be dead by the time he got back to the village. But he still hunted with the rest—probably to take his mind off it, preferring to leave his wife alone with the sick child in her lap and the long expressionless vigil which waits for death.

As he told me about the night, he followed automatically in the tracks of the others who had set out before dawn. Now the strange, pale light of early morning came smokily through the trees like an English autumn. A night bird hooted close by. Kumseran called, his voice suddenly thin and piping to make it carry. The night bird appeared in the form of a young man who had been investigating some *paca* tracks down by the river, but with no success. Strange to think that I was out hunting with 'Red Indians'. It felt much more like being out with a party of nineteenth-century frontiersmen, waiting all the while for the Indians to appear. Appear they did, when we emerged on to the savannah and found them grouped under a spreading cashew tree. The deliciously juicy fruits had all been eaten and the ground was littered with the nuts that attach them to the boughs, looking as if somebody had emptied a tin of those salted shapes so popular in Europe.

We entered the forest again, looking for tapir. Suddenly one of the dogs darted into a thicket and bagged an armadillo. The whole pack was after him immediately, chasing him round and round in circles, trying to wrench away his prize. The dog hung on grimly, bounding into the air, avoiding the men who were also trying to seize him. At last one of the villagers got him and held him up by the scruff of the neck. The other dogs surrounded the pair of them, leaping and snarling. I was the only man wearing boots, so I waded in, kicking the dogs away as best I could, but they scrambled up again and closed around us, crazy with hunger and impervious to shouts and kicks. The Sherente, on the other hand, were not going to surrender the meat, so the battle raged in deadly earnest. We managed to tear the animal out of the dog's tiring jaws and to disperse the others with kicks and blows from the flat of our machetes.

The little armadillo seemed a pathetic trophy after all that. Its plates were gaping and bloodied, like a frigate hit amidships by a large mine. I doubted if it would feed a single family. I felt sorry too for the dogs who yelped and squealed in the under-growth. They were lean and stringy, chronically underfed till their bones knobbled their hind quarters and looked as if they would burst through their taut flesh. They lived by scrounging desperately around the village, competing with the children for fallen scraps in a life-or-death struggle which was heavily weighted against them, since they would be driven off with kicks or stones whenever they became too importunate. The children used the puppies as playthings, ignoring their plaintive mewing. The men used the half-grown, half-wild curs as hunting dogs, convinced that they would be more efficient in the chase if the spur of hunger were added to their natural savagery.

They ran down little game because there was increasingly less and less game to run down in the regions where the Gurgulho men hunted. At intervals along the Tocantins, where the babassu palms stood thickest, settlers had established their precarious footholds in the jungle. They had driven back the trees with fire and loosed their ungainly, humpbacked cattle to roam the land in search of pasture. The cattle frightened away the deer and ate the sweet grass from out of the mouths of the tapir. If an occasional jaguar or ocelot took the opportunity to live easily off these slow-moving and incautious meat reserves, then sooner or later it would be tracked down. The backwoodsmen, colour of burnt caramel, with their ancient ·22 rifles and specially imported hunting dogs, would run them to earth and kill them or die in the attempt. But in the end there were more settlers than jaguar. The game moved away westwards just as the Shavante had done a hundred years before, and the land lay dead and empty under the yoke of progress.

The Sherente knew this, but they refused to accept it. Just as a man faced with catastrophe lives on the bromide of routine, so they hunted where there was nothing to hunt, because to abandon hunting would be to give up their claim to be themselves and to lose their self-respect. Day after hot day they toiled through the hinterland away from the Old Water, as they called the Tocantins, harrying the few remaining tapir, digging the armadillos out of their burrows, surprising an occasional deer.

We hunted through rocky hillocks and stunted trees that were big enough to obstruct yet too small for shade. We never did get that tapir. We gathered in mid-afternoon and rested. Even the Indians were tired. The sweat was like oil on our bodies, in the palms of our hands, over the backs of them. I wiped my face Indian fashion, using the sharp edges of blades of grass to shave off the perspiration. Still no water. We came to some marsh water, so filthy that I could not face it. Later we followed a series of small, choked pools which would be a creek when the rains came. The last few miles might have been the antechamber of hell. We were approaching the Tocantins and the ground had been 'cleared' for cultivation by the slash-and-burn method used by settlers and Indians alike. We trudged through a carpet of ash. The small, charred spikes of what were once trees presented their barren teeth to the sky.

But the Gurgulho, when we came to it, was as rainbow clear as ever. Women were splashing about in it, and they withdrew hastily when they heard the gruff shouts of the returning hunters. We flung ourselves into the soft water. We were home.

Half an hour later Jacinto and I wandered back to the hut. It was beginning to get cooler, and the slow life of the village uncoiled for the evening meal. We were not greeted. We brought back no meat, so there was nothing to be said. Jacinto disappeared to help himself from the store of food I knew he kept hidden in his hut. I had not even the energy to go to some of the reserves which he, like everybody else, knew that we possessed. I looked at our rucksacks with distaste. They had been imperfectly laced up and I knew without looking into them that the women had once again been through all our effects. Not that they took anything from us. They simply made a mental inventory of everything we possessed and then devoted their lives to begging these items from us. These were the scissors, knives and few bolts of cloth we held in reserve to give to those who performed special services for us, and we could not afford to give them to any and every importunate harridan. Nor did we feel inclined to part with our few tins of dried milk or cocoa. We had only been a week or so in the village and already the memory of the down payment for our food was beginning to grow hazy. Jacinto had taken to ostentatiously not eating at 'mealtimes', and his women were more and more inclined to say that they had no food for us. What food there was used to be served to us as if we were eating the final rations of a beleaguered garrison; after which all the children of Jacinto's and the neighbouring households would crowd round us and watch it into our mouths, a technique which often had the desired effect in that Pia found herself unable to eat at all and gave her portion away.

Where was Pia anyway? I went from hut to hut, asking after her in my rudimentary Sherente. More laughter at my expense. Pia was always so busy, gossiping with the women, going on jaunts with them into the bush, ministering to their sick children, that I could rarely find her at any given moment. This delighted the Sherente. They would never bother to look for their own women folk and usually did not need to, for they stayed at home; so they found it especially comical that I should be ever seeking her like the cuckold in a French farce.

I found her at last behind Sizapi's hut, sharing a basket of buriti fruits with the women of the household. Now, I thought, I shall have a sympathetic listener to hear about the hard day I have had. I told her of the trials of the hunt, striving to create an impression of heroic modesty, of additional sufferings nobly borne but omitted in the retelling. She was as solicitous as I had hoped and it was delicious to lie in my hammock and bask in her concern. But a moment later she shattered my self image. I asked her what she had been doing all day and she told me she had been out with a party of women. They had been over to a certain creek, which I knew to be a good seven miles away, and had returned in the early afternoon, their baskets laden with wild fruit. She showed me the carrying basket she had been given and told me how she had brought it back slung from her head after the fashion of the Sherente. It had been a heavy load and she now had a stiff neck. Since I had brought no meat, supper that night consisted of Pia's fruit together with some rice which Jacinto's wife produced with very bad grace.

Both of us ate ravenously, ignoring the children who gathered to watch the performance. I could feel my stomach distending as I forced more and more food into it. It was a habit we had learned since our arrival. When there is food, eat as much as you can. You never know when you will eat again. A couple of lean days had persuaded us of the truth of this unspoken aphorism. Today we had the sensation which the Sherente cherish and which is much celebrated in their stories: the pleasure of feeling our bellies grow big with food. The basket of fruit was nearly empty and the nervousness of the children was increasing when we heard the first wails coming from Kumseran's hut. A man's voice, hoarse and quavering, rising and falling in cadences which were strangely musical, singing perhaps in a new dimension, along a scale of which we were ignorant.

We half ran towards the noise, but checked ourselves when we observed the villagers quietly going about their business. Once more I was covered with confusion and unwilling to intrude my curiosity into a dead girl's house. But Pia had tended her. She hastened on to enquire after her patient, while I hung about uneasily outside, listening to the melancholy keening.

A moment later two children burst out of the hut and stopped short when they saw me in the dusk. One of them scuttled back in again to announce my arrival and a hearty voice called out. 'Come in, Wakuke, come in.'

The hut stank of rotting fish and diarrhœa. There was no lamp, for the family had long since used up its supply of kerosene in their vigils over the sick child. In the firelight I could just distinguish the two women sitting against the wall with the child on their laps. Pia was bending over her. Kumseran sat with his head in his hands and wept. One of his sons was eating unconcernedly. The other was stretched out on the platform bed, apparently asleep, but he scrambled into a sitting position when he heard me come in. They pushed over a log and I sat down. The wailing filled the room, flaring and subsiding like the fire in our midst. I felt sick.

A moment later Pia asked me to fetch the veganin and I left the hut with relief, gulping in the night air to take away the taste of sickness and putrefaction. At least the girl was alive. If Pia wanted veganin, then the girl must be alive. But then why was Kumseran wailing? Outside Jacinto's house the evening crowd was beginning to gather.

'Wakuke,' they called, 'come and sing. Come and play your guitar for us.'

'Wait!' I replied, with a whining inflection of exasperation. 'I am going to fetch roots (medicine) for the child.'

I felt ludicrously pleased with myself that I was able to bring out the sentence in Sherente, without stopping to think. I was tired and light-headed and my feet hurt. I kept thinking incongruously all the way back to Kumseran's hut, 'Now I am getting somewhere with these people,' and I felt guilty that I did not care more about the dying girl.

'Wakuke,' I was pursued by the voices. 'Wakuke, come and sing. The people are waiting.'

'I am sleepy,' I replied. This is a laudable condition among the Sherente. They would have laughed at me had I said I was tired. 'I will not sing tonight. Kumseran is sad.'

They laughed at me anyway. Jacinto laughed until the tears came into his eyes.

'Kumseran will not mind whether you sing or not,' he said, and the others got the giggles all over again at such an incongruous juxtaposition of ideas.

So I sat down and gave them a spirited rendering of the latest carnival tune from Rio.

Kumseran's daughter did not die, but no sooner had she passed the peak of her fever than there was an epidemic among the smaller babies. From one day to the next it seemed that every hut had its corner of sickness: a silent woman brooding like an icon over the baby in her lap. The infants appeared to be melting as they lay, shrivelled and vomiting, in the rags which swaddled them.

Sizapi was particularly worried about it, since he was in the process of building up his own reputation as a seer.

'There are only two *sekwa* who are stronger than I am,' he once told me, with characteristic lack of modesty, 'and they are old. Now it is I who have to look after the children.'

As a young man he had studied under a seer so powerful that he could silence the thunder, drive away the rain and provoke a drought at will. The old man had been a benefactor of his people. He used to say, 'Go to such and such a place and you will find pig,' and the young men would go out and find herds of wild pig, so many that they could hardly carry home the meat when they had hunted. When they came back they ate until they got bellies like women, and they danced and sang because their seers were so powerful that they could protect them from sickness and want.

'But he could kill men too, if he wanted to, couldn't he?' I asked maliciously.

Sizapi parried the question.

'He was a good man. *Sekwa* are good men who look over their people and make them strong.'

I knew from the histories that the seers did look over their people, but I also knew that the people did not always like it. In fact they went in terror of the powers which might be wielded by an unscrupulous magician. So much so that the calling became an extremely risky one. Sooner or later a seer was likely to be done to death by his own people, at the orders of their chief, and his body specially disposed of—a stake driven through his heart, the palms of his hands and the soles of his feet incised and rubbed with charcoal. But inevitably, when the people were faced with a new catastrophe, they would be obliged to seek out another magician and invoke his aid.

Sizapi had set himself up as the special protector of the children. Once in a while he had visions in which the Sun came to him in human form and instructed him to perform certain dances and sing certain songs. Then he would gather the people together and lead them through the steps, always ahead of them in the moonlight like the Pied Piper, the swishing beat of his dance rattle lulling them in its rhythm, sleep dancers stamping and jumping, two steps forward, one pace back, till the stars reeled out of the sky and the first light of day sent them stumbling to their huts. In this way he channelled the powers of the community on behalf of their children.

The epidemic was a personal challenge to him. If the babies died, his reputation would suffer. Worse still, it might be whispered that Sizapi's powers had curdled within him, that he was indeed responsible for the blight upon his charges. So he fasted and walked alone in the forest, hoping for supernatural guidance.

A yellow haze hung over the land. The rains were delayed in their coming and the thunder stayed like a bruise on the skyline, gloomily menacing. The ground began to crack and even the rank steppe-grass shrivelled up and died out on the savannah. The men did not hunt so often now, and only a listless few would wander out into the bush in answer to the early morning halloo. For the most part they sat about, winding their coils of tobacco from one great spool to another or lay drugged with the opiate of idleness. The women went out scavenging and returned with laden baskets to feed their families. Sickness and suspicion clogged the heavy air.

I saw even less of Pia. All day she was away, mixing portions of milk for the children, giving them minute doses of medicine to try to stop the fever and the perpetual diarrhœa. The women watched her passively as if she was performing a rite. She begged them to force nourishment into their skinny charges, but they shrugged their shoulders. What was the use, if they would not eat?

I spent much of my time with Sizapi, improving my Sherente by hearing about his dreams. He was dreaming vividly about this time, voraciously but all to no purpose. One day he surprised me by telling me that he had been on a visit to 'Jesus, Our Mother'. That is, he would have visited him (or was it her?) had a long-tailed monster not turned him back when he reached a log leading over an impassible swamp. I questioned him as best I could about this polymorph and could only conclude that the confusion of sex had been aroused in his mind by hearing Catholics talking about mysterious figures like 'Our Lady of Jesus'. In any case it was clear that he was getting desperate, since the Sherente deity Waptokwa would not respond to his prayers.

I was getting worried too. My Sherente was improving so that I was able to understand the gist of conversations which were often carried on in the belief that I was to all intents and purposes a deaf witness. It was clear that the people were taking a serious view of these illnesses. The ominous word 'sorcery' kept coming up again and again. An outsider would be more than likely to be accused of such malevolence. Would not the co-incidence of the epidemic with our presence in the village be taken as circumstantial evidence of our responsibility? Especially since Pia was so actively in contact with the sick children? I did not tell her of my suspicions, since there was nothing to be gained by her leaving off the rudimentary treatment she was able to give to the babies. But it seemed to me that Jacinto was more shifty-eyed than usual, and I felt that some of the men were avoiding me. Only Sizapi, when he was not out in the jungle, appeared to welcome my company. Perhaps both of us were on trial.

One day Sizapi decided to act. In the evening he addressed the group outside Jacinto's house, but only conversationally, since he was a poor speaker and did not feel he had the seniority necessary to qualify him for the full-scale oratory which is at once the mark and the intrument of prestige. The following day, he said, the people were to prepare themselves for a ceremony. He would dance for them during the night when the moon was up, and when the first rays of the sun came out of the east he would look over all the children and they would be strong.

The announcement was received calmly. Something of the sort had been expected. The people set about collecting the latex to glue stripes of charcoal on to their bodies in the requisite ceremonial patterns. Those of them who were lucky enough to have some of the precious urucu seeds which gave the bold red colour essential to traditional Sherente designs hid them care-fully lest their kinsmen should beg for them and they be forced to choose between the shame of refusing a close relative and the ignominy of appearing without the attractive paint.

When the sun was high Wakuke arrived. He must have got wind of the ceremony somehow, perhaps a messenger was even sent after him. The men stopped whatever they were doing. Even those who were lying inert on their sleeping mats seemed to take on the muscular reflexes of expectancy. A hoarse, joyful carolling could be heard the other side of the ceretona trees.

'Wakuke,' they murmured to each other. 'Wawen', giving him the honorific title of Old Man.

I had not anticipated anything like the figure who came stumping in at the far end of the village. Wakuke was short and well muscled. He carried a long sword-shaped club carelessly over one shoulder and walked with a slight swagger, like a poacher returning from a good night out. He was singing, or

Carolina: The port

Tocantins River with motor
boat in the dry season
(*above*)

Pia with Jacinto's second
wife, Garricha, in the background (*right*)

Sherente: Pia receiving the
name Tpedi

rather shouting, at the top of his voice, a performance which was clearly much appreciated by the pair of mongrel dogs who followed at his heels. As he came nearer I could see his wicked little hook-nosed face, reminiscent of Mr Punch, and his gap-toothed smile. He was in the best of spirits. I found out that he usually was.

He marched up to where I was sitting with Jacinto and un-slung his club. Taking it in both hands he lunged forward towards us as if he were about to bayonet us and delivered him-self of a hiccoughing rush of words. They skittered unintelligibly down the scale of his speech until he caught them up sharply in what would have been a magnificent final belch, an exclamation mark straight from his solar plexus, but he cut it off sharply. There was a moment's silence while his mouth worked, a stammerer wrestling with his tongue, and then he rounded it off with an impressive click that made me jump despite myself. That was only the first paragraph. Swinging on to his back foot he lunged at us again and went on with his oration. Pia appeared, round-eyed as the children who were with her, to see what was going on. The men who were with me sat about elaborately un-concerned. Jacinto looked at the ground. I did not know where to look. Wakuke gave his final click, then he turned to me with a grin and said *'Como vai'* in good Portuguese, following it up with a request for tobacco. I had none to give him, but I managed to buy some from one of the others and make the old man a present of it. He said 'Thank you', again in Portuguese, and I realized that he was the first Sherente who had ever thanked me for a gift.

'They told me you had come,' he began, after he had lit his home-made pipe. 'I could not come to see you before because I have so much work to do.'

'What are you so busy with at the moment?'

'Clearing my gardens for next year. And building a new house.'

'Do you have anyone to help you?'

Wakuke chuckled.

'They help,' he said, 'by lying in the shade when it is hot. And at this time of year it is always hot. Then they are hungry and they have to go out and hunt for food. No, I do most of it myself.'

The thought seemed to amuse him immensely. I asked him why he lived so far from the village. He considered for a moment before replying.

'There is too much gossip in the village and anyway I want to raise pigs.'

The reasons seemed rather inconsequential, especially since there were plenty of pigs in the village. They rootled around in the surrounding bush, living mainly on human excrement, for which they had developed such a taste that one often had to fight them off in order to be allowed to excrete in peace. But I had not time to press the point. Other men began to crowd in and to swap news with Wakuke.

Pia and I excused ourselves and went for a swim. On our return I noticed that old Waerokran was thatching his house and, feeling that obscure sense of encyclopædic obligation from which anthropologists suffer, I went over to observe how he did it. He was pleased by my attention and beamed at me with his gummy witch's grin. I found him and his ménage quite revolting. They were also pitiable, but somehow my pity never quite managed to cancel out my feelings of revulsion. I think he must have been the oldest man in the village. He was scrawny and silly, though I suspect that this was due to his nature rather than to his senility. Late in life he had married again, a nondescript negroid woman with spiky hair and goitre like a beehive on her neck. She had been raised among the Kraho Indians and spoke more of their tongue than she did Sherente. Their daughter, now about nine, was deaf and dumb from birth. They also had a well-formed baby son, still suckling in his second year. The sight of the child in their hut, which was impoverished even by Sherente standards, with his parents leering like gargoyles above him always filled me with unease. The sister acted as maid of all work or more nearly as beast of burden in Jacinto's household. They fed her when they could not avoid it and they had given her one sack-like dress in return for her labours, a piece of generosity which Jacinto's first wife never stopped talking about. She was also beaten occasionally because she stole, or they said she did. Certainly she was a fiddly child who rootled in everything, perhaps compensating through her touch for the faculties which were denied her. I always thought that Waerokran and his family were living reminders of the blindness and cruelty of fate.

Today the mother and her children were scuffling like mice in the patch of shade provided by half a roof. I could hear the wordless chirruping of the girl and the bad-tempered expostulations of her mother. Was there no tenderness in the woman at all? Not even for her baby? She never showed any. Whereas the old man positively dribbled with pathetic pride when he looked at his son.

Waerokran and I straddled the roof, bowed under the burden of the sun, while I watched his deft fingers hooking the ends of the palm fronds and weaving them into the thatch. His wife grew more querulous as the afternoon wore on. She appeared to be working herself into a fury of petulance about something or other. Now there were thumpings and noises like an animal in pain. I peered over the edge of the roof at the same instant as two indistinguishable bodies, like lizards copulating, threshed out of the hut into the sand. They separated for an instant into mother and daughter. The child tried to scramble to her feet but her mother pushed her down and whacked her again with a leather thong, lost her balance, fell on top of the squirming girl, clawing at her.

I was horrified and paralysed. My instinct to leap down if necessary on to the back of the foaming woman, and intervene was neutralized by a myriad confused thoughts. Was it my business? Nobody was taking any notice of the scene, so why should I? Would it be resented if I did? How would it affect my position in the village? Ought not one of the elders to intervene? Ought not Waerokran himself to do something? I looked hopefully towards him, but he was smiling sheepishly at me as if asking me to excuse him for his family's behaviour. The woman dodged back into the hut and out again with surprising agility, carring a machete. She caught the scampering child in two strides and gave her a blow with the flat of the blade. I started to slide down the roof, determined that this had gone far enough. But at that moment a whirlwind erupted from Jacinto's hut. Pia came racing over, running in great bounds like an athlete. She was white with fury. She caught hold of the frenzied woman and shoved her roughly to one side, so that she lost her balance and rolled over on the ground. The child scuttled away without a backward look. The two women faced each other, Pia babbling angry remonstrances in incoherent Portuguese while the madwoman, calmer now than her aggressor, came out with a string of disjointed explanations.

I suddenly felt futile and as ineffectual as a stray man in a maternity ward. Pia turned to me, the anger still taut about her lips.

'I was just going to try and stop them,' was all I could find to say.

Later on I took out my feelings of inadequacy on Jacinto. He was the chief, wasn't he? Why didn't he see to it that such things did not take place in his village? And if they did, surely it was his business to come out and put a stop to them, instead of skulking in his hut and yarning like an old man. This riled him, as I knew it would. He told me in his whiny voice that he was always telling his people how they must behave. He told them until he was tired of telling them, but they would not listen.

That was perfectly obvious, I retorted (I was determined to rub it in). His own daughter's stepdaughter was persistently maltreated in the very next hut and he never did a thing about it.

'But that is because she has no mother,' replied Jacinto, somewhat surprised. The obvious, cruel finality of the explanation took my breath away. I let the subject drop.

That night Pia had a high fever. At first I thought that she had been over-excited by the afternoon's episode and by her subsequent feelings of confusion, tinged with shame. But at dusk she was running a high temperature and it was clear that she was really ill. I settled her in her hammock, feeling her hot skin already flaky to my touch. I put both our blankets over her, but they could do little to thaw out the fever which by now was frozen into her bones. She lay shivering and drenched with sweat, while the villagers came one after the other to peer at her and to repeat that she was hot, very hot, taking a certain pride in the very excess of her ague.

Many of them were already painted for Sizapi's festival; the black, angular designs like lattice work on their bare torsos. The women too had bared their breasts and allowed their men to paint them. Only one Brazilian girl married into the village had refused, snickering, and been backed up in her refusal by Jacinto's senior wife who likewise declined to expose herself in this barbaric fashion. His second wife, Garricha, was painted like the rest in order to identify with her husband's people, but with her wavy hair, straight nose and flaring nostrils she made an incongruous Indian. I realized what an effort it must have cost her to discard the ingrained prudery of the backwoods in deference to Sherente custom. She was young and strong and looked splendid as she stood, with an air of abashed defiance, in the long line of Indian women.

Sizapi wore a palm circlet around his head and his body was plastered with fine white down. The last rays of the sun glinted on the scarlet stripes of his regalia. There was a new assurance in his manner as he walked up and down, marshalling the lines of dancers and motioning them into position with his painted dance rattle. Only the gentle, diffident voice cajoled rather than led, questioned rather than inspired confidence.

Wakuke had already joined hands with a group of men and was leading them in a rollicking jumping dance like a chain of frogs. The villagers laughed uproariously at him, and the more they laughed the more he clowned for their benefit.

Sizapi took the opportunity to leave the scene of buffoonery and go and see how Pia was doing. He stooped under my hammock ropes and stood for a long time looking down into her flushed face. Then he said quietly: 'She has been bewitched.' Suddenly I saw her illness in a new perspective. Up till then it had been a matter of aspirins and of battle against her soaring temperature. Now it was a matter of public concern. She had been bewitched! My first thought was one of relief. At any rate that let her out. If she was bewitched she could not be the source of the witchcraft, unless the situation was more complex than I realized.

Sizapi instructed her to get up, so that he could remove the poison from her body, but Pia just lay there with fluttering eyelids, hearing nothing that he said to her. I bent over her and explained the situation.

'Must I really get up?' she groaned. 'I don't really think I can.'

'We cannot refuse,' I told her. 'Come on, I'll help you.'

I disengaged her from the hammock, and she stood unsteadily against me, her teeth chattering. Sizapi looked hard at her. From outside we could hear the shouts of Wakuke and the men who were apparently still making whoopee. Pia staggered a little and asked to go back to bed. But Sizapi had not finished. He spat on her forehead and I saw her recoil in surprise. Then with deliberate movements he rubbed the spittle over her brow, kneading it into her temples as if he was dunning a lesson into a stupid child. He slapped her lightly on each cheek, pulled open her shirt and spat on her breast, once more rubbing in the spittle towards the points of her shoulders. Pia looked as if she was about to vomit. Now Sizapi was working up to the climax of the rite. He shook his dance rattle over her and then turned away, his body tense with the effort, to push the unseen miasma out of the door. Three times he tore the poisons out of her body and dissipated them into the open. Then he relaxed, tired but confident, and Pia was allowed to sag back into her hammock.

'Now she will not die,' Sizapi told me conversationally. 'I have taken out Estiva's witchcraft.'

Estiva's witchcraft. Estiva, the wife of Waerokran, that dark, knotted figure—she looked the part. She was also an outsider, from the Kraho, and who knows what strange powers they disposed of? And the very day she was attacked by Pia—her attacker was stricken by a raging fever. But . . . a witch. Did they still kill witches, I wondered. I ought to do something to prevent it, but what? And all the time there was a corner of my mind which kept reiterating—if it happens, I only hope I don't see it.

It was quite dark now and there was utter silence outside. I could hear only Pia's stertorous breathing and what might have been the shuffling of bare feet as the dancers took up their positions. I went out to observe the ceremonies, but as I passed through the assembly room I was brought up sharply by a dark shadow crouching on the floor. Did its eyes really glow? Or is it only memory which redraws that instant for me, using emotions for colours? Those eyes glow still in my mind, like those of an ownerless cur which once woke me from my sleep as a child and sent me into hysterics.

I stood quite still and brought out automatically the formal Sherente sentence of enquiry—'What do you seek?'

Estiva got up from where she was sitting.

'I came to see how your woman was,' she said, and there was real fear in her voice. I realized that her life depended on Pia's progress. 'I wish to see her, for I am full of shame.'

I went back into our cubbyhole and lit the kerosene wick. But before I could say anything to Pia about the visitation, Estiva appeared in the doorway, dishevelled and staring, the shadows playing horribly on her distorted features.

Pia screamed, only it wasn't a scream. It was an abortive, nightmare noise, truncated in its utterance.

'Estiva has come to see you and to say she is sorry,' I think I said, appalled by the nannyish futility of the words.

The singing started. Sizapi's high melodious voice chanted a few bars and then men took it up, grunting every other note with breathy emphasis as they stamped out the measure.

Estiva started trembling. For the first time I saw her with real pity. I searched for some phrase of comfort, but was lost in the tangle of an unfamiliar language and a situation which I did not fully understand. Anyway, she disappeared, muttering excuses as she ran.

I followed her, but my hand was seized in the darkness before my eyes had adjusted to it and I found myself whirled into the business of the dance. Backwards and forwards we stamped, while I strained to catch the words. Every now and then there would be a break and I would slip inside to see how Pia was. She slept heavily and I was obliged to dance on till I lost all notion of time and space, since I could not read the stars.

Each time there was a lull in the dancing I sneaked back into the hut to see how Pia was, only to be called out again and to make my stumbling way back to the dancing ground over the huddled bodies of the women who had brought their sleeping mats to the edge of the plaza. As morning approached I felt exhausted, but mentally rather than physically. I felt like an acolyte in ceremonies which I only dimly understood and which I was getting too tired even to observe with the proper detachment. More than anything else I wanted to be allowed to stay in my hammock and sort out my impressions. But this time it was Wakuke himself who came to fetch me.

As we stepped into the cold starlight the singing stopped and the dancers squatted down around the fires. Nobody spoke. They sat motionless, some of them still panting from their exertions in the dance. I was surprised, for Sherente are usually casual and conversational in the midst of even the most important cere-monial. We sat there for an hour, or perhaps it was only for a few minutes. Even the fires stopped crackling and glowed like wounds in the night. A sighing movement ruffled through the lines of men. Wakuke nudged me.

'There it is. Listen!'

I listened.

'Can you hear it?'

'Hear what?'

'The souls of the dead. Listen, they are whistling around us.'

All I could hear was a dull humming noise, probably coming from inside my own head. Suddenly Sizapi jumped up and shook his dance rattle towards the heavens. The men all bounded to their feet, and the thudding, bellowing dance broke out again.

Just before dawn I fell asleep, to awake confused and unable to collect my thoughts. The sun was already clear of the horizon. Pia was asleep. I could hear a murmuring from the other side of the village, interrupted by the staccato sounds of oratory. I realized that a meeting of some sort was being held and further-more that it was no ordinary meeting, otherwise it would have been convened in the middle of the village. I hurried into the open, where the women were still lolling, suckling their children

THE SAVAGE AND THE INNOCENT

where they had slept. There was not a man in sight. From the far side of Sizapi's hut I could hear Wakuke making a speech. All the men appeared to be there, sitting sullenly in congregation. They did not look up when I arrived nor give any indication that they had noticed my presence. I felt like a ghost. It seemed as if they deliberately mumbled and jumbled their words to make it harder for me to catch the drift of their discussion, or perhaps I merely suffered from woolly wits and lack of sleep.

Wakuke for once was deadly serious and he was clearly arguing a case before (or was it against?) the rest of the village. He was talking about a piece of wood. Somebody had taken it or stolen it. A woman. Killing. Kill. Don't kill. Sickness and death. Angry mutters from his audience.

'Fetch it!' commanded Wakuke at the climax of his peroration. He pointed with his club at a ten-year-old boy who huddled against his father for protection.

'Fetch it!' thundered Wakuke again.

The boy did not move. There were expostulations from the crowd. Two other men started to make speeches simultaneously. Wakuke sat down. I sidled over to him and he grinned at me. 'Give me some tobacco,' he said, with leisurely unconcern.

'What are they saying,' I asked him in Portuguese, to make sure I understood the answer.

'They are very angry,' he replied unhelpfully. I could not get him to tell me any more. As the meeting broke up he remarked that the woman would not die this time.

'Which woman?' I asked.

'Your woman,' he replied to my horror.

The day was heavy and somnolent. Wakuke went back home. The villagers slept. Pia lay in the grip of her fever. Jacinto and his family avoided me and failed to provide any food. I went to seek out Sizapi and found him too exhausted to speak. Then, on an impulse, I went to look for Estiva. Her half-thatched hut was empty, denuded even of the rags and bundles which made up the family's belongings. When I enquired after her, the women pointed their lips indifferently towards the jungle. They had gone to pick babassu nuts, I was told. They might all have been killed during the night, for all I knew. Once again I was aware of how ignorant I was concerning these people. Not only of their language, but of their actions and more especially of their motives. I had been scribbling notes on every detail of Sizapi's ceremony, trying to piece together his visions with the snatches of explanation I had been proffered, trying to make out of the epidemic, the dancing and the witch hunt a single coherent whole. I had even believed that their insistence on my taking part in the ceremonial implied a degree of acceptance of my presence and my person. But did it? Pia might have been killed while I was out there playing Indian. They might still intend to kill her. I did not know. I felt I knew nothing—at any rate, nothing worth knowing—and was overcome by a profound depression to which I would have abandoned myself had it not been for the necessity to look after my patient.

For two whole days I went about like somebody under a taboo, isolated in the midst of indifferent strangers. Then I decided that I had to act to restore my self-confidence otherwise we could just as well finish our research there and then. Besides the village badly needed a new topic of conversation. So I instructed Jacinto to call a meeting of the elders that night.

Pia had recovered sufficiently to attend it with me. We sat as usual on the left of Jacinto while the men gathered. The whole scene was etched sharply in the light of an early moon, but tonight we were unreceptive to such impressions. Instead I could not help noticing how absurd the old men looked in their shirts and shorts and being irritated by their pointed gossip.

'No. I have no grain left. None at all. Can't think what we are going to eat tomorrow. But then we did not eat today either.'

'That's right. We are hungry too. I would go out hunting if I had any shot left, but I used it all up long ago and it is dear as the devil in Tocantinia.'

Jacinto said nothing. He was looking both guilty and unusually nervous, I noticed with delight. There was no doubt about it—he thought I had summoned the council to explain to them that I had paid him to feed me and my wife and that he was in fact near starving us. Such a revelation would not only shame him publicly but would also embarrass him, since he had been putting off his creditors and defaulting on his chiefly obligations of largesse by explaining to all and sundry what a drain we were on his household and describing in lurid detail the style in which he kept us.

I wished that I could make a speech like Wakuke. I sensed that a few impressive rhetorical barks would do my prestige more good even than a cocked hat and a sword, but my Sherente was not up to it. I spoke in Portuguese, so that I would at least have them at this disadvantage.

I told them that I had been observing their people and that I sympathized with them in their plight. I had noticed that they fought sickness without medicines, that they hunted without powder and shot, that they cleared their gardens without proper tools (cheers). I had seen them go hungry (loud and prolonged cheers) and had felt the pangs of hunger as one of them (a meaning look in Jacinto's direction, which he avoided). I had therefore decided that I would make the village a present of a steer. They could slaughter it and get big bellies and have food enough for the name-giving festival about which I had been told so much. We too would share in the banquet which would make the people strong for the harvest season. Then there would be no more hunger, since the villagers had all worked so hard and planted so much (no cheers here) and there would be no more sickness either. I did not know how to add 'and we will all live happily ever after.'

The news was received in the customary silence. Finally Dbakro said that they would be happy to eat my steer. He hoped that it would be the first of many and that when I went back to Rio I would get the government to send many presents to the Sherente. I was a good man and a true friend of the Sherente, he added, and he hoped that I would come back and visit them again, bringing with me many presents such as the Indians really needed, not just a few old knives as I had done this time.

Jacinto, reprieved, began throwing his weight around. He cross-questioned me about the size of the animal, where I would get it, how many days would elapse before it arrived and so on. Then he launched into a discussion of the necessary dispositions for its disposal. For the rest of the evening they discussed the price of cattle with all the academic precision of connoisseurs who had never owned a single head between them. I felt certain that, if they were contemplating any move against us, then they would wait until after they had eaten the beef.

Next day Pia was well enough to look after herself so I did what I had been wanting to do for a long time. I went to visit Wakuke. I was escorted by a nephew of his who showed me the way southwards to where the old man lived. It was mid morning when we arrived. Just as the trail seemed to get lost in the thorn bushes we found ourselves up to the knees in Wakuke's dogs. There were a couple of lean-tos, each big enough for two people but no more, built of ferns. The clusters of gourd water-bottles nestling round them in the shade showed that they were occupied, but there was no sign of the old man. We found him a minute later with his two women in a grassy glade where they were working on a long, thatched house that looked like the Sherente equivalent of a banqueting hall. Inside the uncompleted structure lay a strapping young Sherente, a distant kinsman of Wakuke's, presumably gathering his strength before coming out to help.

Wakuke slid down and greeted me without any of the sly reserve which usually characterizes Sherente overtures to white men. We lay down together on a soiled sleeping mat and immediately his dogs wandered all over us, waiting to be petted. I produced a length of tobacco for him and some small gifts to loosen his tongue. Then we settled down to talk until, as they say in the interior, the sun starts to hang.

He told me proudly that his full Portuguese name was João Pedro da Silva Cavalcante Capador and that everybody up and down the river knew him at least by repute. He had been a friend of the white man ever since the times when the first missionaries had been carried into Goiás on the backs of friendly Indians and had terrified the children with their cavernous sunken eyes and bushy beards. He described for me graphically how he had wept when as a child he had seen the first missionary, convinced that he was a *romsiwamnarin* monster. This was a piece of poetic licence, for the missionary in question had been sent to the Sherente long before Wakuke was born. I guessed that he was repeating the story word for word as his grandfather might have told it to him. Either that or he was poking subtle fun at my own bearded face. But he was a good raconteur and it was all I could do to keep my critical faculties awake as he talked. I wanted to drowse and let my imagination meander through the times he talked about, trying to picture the Sherente of half a century·ago and how they had got on with the tough and impoverished settlers like the da Silvas and the Cavalcantes who filled Wakuke's tales. Instead I had to swat away the clouds of insects and make mnemonic notes in my ugly little books, trying to reduce the rich narrative to headings and sub-headings as the afternoon wore on. Now and again I needled a remark into the conversation, trying to bring the talk round to more topical affairs so that I could reconcile the indigestible mass of myth and history with the actualities of Sherente life as I partly knew them. And all the time I kept reminding myself that there was an ulterior motive in my visit. I had to get Wakuke to tell me about the meeting which took place the morning after Sizapi's festival. I had to find out just where we stood with the people of the Gurgulho. But Wakuke was not much interested in telling me about the present. The present was a bad time for the Sherente and he preferred not to dwell on it. He was too shrewd to blame all the misfortunes of his people on the whites. Many whites cheated the Sherente and worked against them, he admitted, but then the Sherente made things worse for themselves by their inability to live together in peace or to present a united front to the outside world. All the Sherente villages were intriguing against each other, he insisted, crossing his fingers like swords and rubbing them against each other to illustrate the point. Even within each village there was no freedom from gossip and thieving and fighting. That's why he would not live in the Gurgulho any more. Here by himself he had peace and he could raise pigs to earn a little money, without having them killed or stolen as they would inevitably be in the village.

I told him I knew something of the shortcomings of his people and added that I was inclined to take a very grave view of the fact that they had murdered Estiva on suspicion of witchcraft.

This shocked him into forgetting some of his diplomatic caution.

'But they said they would not kill her,' he stammered uncertainly.

'Well, they have,' I pressed home my advantage. 'Nobody will tell me where she is gone to. I know that she was killed on the night of Sizapi's festival.'

Wakuke was visibly agitated. Before he could turn for corroboration of my story to the boy who had come out from the village with me, I ventured another shot in the dark.

'And they say that they will kill again if the sickness does not go away.'

Wakuke called his nephew over. The boy came grumpily, reluctant to leave the shade for a moment. He was questioned briefly and curtly. No, the people had not killed Estiva. She had gone away with her husband to pick babassu nuts. No, no babies had died. Yes, the people said it had been a good festival and the sickness would soon pass.

'There you are. . . .' Wakuke turned to me.

'Then it is another witch they will kill,' I persisted. 'The people are angry. I heard them the day after Sizapi's festival.'

At last he told me the full story. Certainly the people had suspected witchcraft, for otherwise why would all the babies have fallen sick at the same time? First one suspect had been named and then another. He, Wakuke, had known it was all nonsense and had told them so, but they were foolish and afraid and they persisted in looking for a scapegoat. They thought I and my wife might have brought them this affliction, but when Pia quarrelled with Estiva and fell ill that same evening, then the circumstantial evidence all pointed one way. People began to remember that Estiva had acted suspiciously in the past. A boy claimed that he had seen her burying a peg of wood in Prakumse's garden a few days before all the young plants in it were destroyed by a pair of armadillos. The same boy had later seen her burying a peg outside one of the huts in the village which now had two sick babies. Sizapi had called the whole village to dance and to invoke the help of their ancestral spirits, but to no avail. The identity of the witch still remained uncertain. So the elders had called a council to decide what ought to be done about Estiva.

'What did Jacinto do about all this?' I asked.

'Nothing,' replied Wakuke with a trace of scorn. 'Jacinto is a quiet man who does not like to give advice.'

That was the council I had interrupted. The majority were for doing away with the witch then and there, but Wakuke had demanded more concrete evidence. He had told the boy to dig up the wooden peg which Estiva was claimed to have buried with such ill effect. But the boy refused, saying that he could not remember the exact spot. Whereupon Wakuke had made some scathing remarks about his honesty and trustworthiness, branded his whole story as malicious fabrication and demanded that the villagers ignore the charge.

'But why did you tell me that my wife would not die?' I asked him.

He was genuinely surprised at my naïvety.

'Sizapi cured her.' He smiled at my bewilderment. 'He danced especially for her, and the souls of the dead came to him. You danced for her too, didn't you? And she was cured, wasn't she?'

I had to admit that she was.

4

When traditions wear thin

The villagers now thought of little else but the steer I had promised them. They rubbed their bellies suggestively whenever they saw me and asked me when I was going to get them the meat. As it happened, I had not given the matter too much thought. Everybody except the Indians had a head or two of cattle in these parts and I had assumed that it would be easy to secure a satisfactory animal whenever we wanted it. But this was far from the case. Cattle are wealth in Goiás, perhaps even more so than money, for there is precious little to buy. They are also a status symbol. A man without cattle is hardly a man at all, so the most impoverished backwoodsmen would not sell, even though the price I was prepared to pay represented more money than they would see in three months.

Finally, through Eduardo's good offices I was able to buy a steer from the Indian agency on behalf of the Gurgulho villagers. It was a sizeable longhorn, though lean like all the local cattle, and it was brought galloping into the village accompanied by a flock of boys hallooing with excitement. All the men rushed out to examine the new arrival as it stood, panting and uncertain, in the ring of bystanders.

'Now you can hold your name-giving ceremony,' I said to Jacinto.

He murmured his assent but avoided my glance. I assumed that this was because he was so busy looking his gift steer, if not in the mouth, at least in every other conceivable crevice.

I went off to get my notebooks, but before I could settle down in conversation Pia burst in to say they were killing the steer. A long rattling bellow confirmed her words. As we ran outside there was a stampede amongst the Sherente. They scuttled in all directions like crabs on a seashore while the wounded animal galloped and plunged about the village. A long machete was buried deep in its breast, the blade emerging in a fountain of blood from its throat. The handle swayed above its head as the beast cavorted in pain. I thought of shooting it but Pia, ever practical, pointed out

70

that in the confusion I might just as easily shoot a Sherente. There was nothing to do but stand and watch as the animal groaned and choked in eddies of blood-flecked spume. The men waited unmoved for it to sink to the ground. Only the children still danced around it, a grotesque *corps de ballet*, laughing and gesticulating as they imitated the thrust which had so signally failed to kill the beast.

At long last it subsided in its pool of blood and the men closed in. Methodically they hacked it to pieces with their machetes. Half an hour later its parts were bubbling in every pot in the village while its stomach still ruminated uselessly among the litter of offal which marked the spot of its butchering.

We ate all day. Pia argued it out with the women and finally came away with huge hanks of liver which we salted for the morrow. The others gorged till they fell asleep and then woke up to gorge some more. The fires burned most of the night so that families could help themselves from the pots, and in every shadowy corner there were sounds of crunching as each individual consumed those private delicacies which he was reluctant to share.

By the following morning it was all gone. Soon after sunrise Wakuke arrived, singing at the top of his voice, to collect his share. It was a measure of his prestige among his fellows that he could afford to leave it to them to keep his share for him.

'When is the name-giving ceremony to be held?' I asked him.

'It is difficult,' Wakuke admitted. 'The people say there is no food for them to eat so that they may not feel strong enough to dance.'

'What about the steer I have just given them?'

'They have eaten that already,' he said with an air of finality. 'Now, if you perhaps could give us another steer . . . '

Fortunately I knew no swear words in Sherente. Besides Wakuke was the nicest of men and I suspected that he had been detailed to approach me on this delicate matter because it was known that I liked and respected him. I told him politely that I could not provide another steer and there the matter rested.

Jacinto tried again later. This time it was he who brought up the name-giving, intimating that his people were still quite willing to put on the ceremony for me if I made it worth their while. I told him that it was a matter of complete indifference to us whether they named their children or not. At first he was disbelieving. As days went by and we showed no further interest in the ceremony, he was a little bit disconcerted. The time was ripe for the naming ceremony and the old men were beginning to mutter that it was a shameful thing that it had not already been held. Clearly Jacinto had hoped that we would act as patrons of the festivities on the grounds that they were being enacted solely for our benefit. Now he was caught between our apparent lack of interest and the urgings of his own people.

It looked as if the Gurgulho had celebrated its last true name-giving. But here we were mistaken. We had underestimated both the subtlety of the Sherente mind and the influence of the traditionalists among them. The traditionalists insisted that the children should be ceremonially named. Those who did not mind whether they were or not hit upon the idea of bestowing a name on Pia. This would please the traditionally inclined by bringing her formally into a correct social relationship with the rest of the villagers and would at the same time place her under some sort of obligation which could later be exploited.

Traditionally, girls had been named by the men's societies. Each society had its own set of names and they transferred them ceremonially to selected pairs of small girls, one from each moiety, whenever they were requested by the parents to do so.

Wakuke was immediately summoned to lead the dancing. He appeared at dusk and stood in the middle of the village, alternately swinging his club and uttering a piercing rallying call. A few men came out to join him. I noticed that they were all painted and that they all carried clubs. They formed a circle with him and broke into a deep chant, pounding their clubs on the ground to the rhythm of their song. An old woman came slowly across to us. She was naked to the waist and painted. She took Pia by the hand and led her out to the men. A little girl who cannot have been more than about eight years old was also led out to be named. The little party went from hut to hut all round the village. They sang before each one while the women executed a small, shuffling dance step, so simple that even Pia required no instruction in it. Night fell as I lounged by Jacinto's hut. The singers were getting hoarse by now but the earth shook with their poundings as if they were determined to supplement their vocal deficiencies by the vigour of their accompaniment.

Dbakro was sitting beside me.

'Now she will be called Tpedi,' he said. The name meant Fish in Sherente.

'Tpedi,' I repeated after him and they all giggled.

'Call her and see if she comes.'

I gave my best imitation of the Sherente way of calling. 'Ooo Tpedi!' the first Ooo being as close as I could make it to the whistle of a steam engine.

'Ooo Wakuke!' came Pia's answering hail. She came over to us to my great relief, for Sherente love to see wives make their husbands look foolish, and Pia's independence had become proverbial.

'Tpedi,' the children called to her, 'there's Tpedi.'

'What do they call you in your country?' asked Kumseran.

'Pia' we told them for the hundredth time. They loved hearing these foreign words and trying them like strange wines on their tongues.

'Pia, Pia,' they repeated. 'What do you call a house in your country, Wakuke?'

I told them.

'And in your country, Tpedi?'

This was a standing joke. They had been fascinated when we told them that Pia and I came from different tribes who spoke different languages and that it was two days' journey from my home to hers. Ever since then they had insisted that we should while away the evenings by giving them comparative vocabularies in English and Danish.

'Hus,' said Pia, and the word was greeted as always with a roar of laughter. The sharp glottal stop in Danish invariably struck them as improbably funny. But Pia could take comfort from the fact that the biggest joke in their eyes was at the expense of my tribe. They had examined our letters minutely when they came down river, scrutinizing the envelopes with a child's eye for insignificant detail, and had finally cross-questioned us about the design of the stamps. We had explained that the man on Pia's letters was the chief of her nation, and the woman, chief of my tribe! Did she get up and give advice like a chief should? Yes, I told them. She often got up and spoke. At this they exploded with laughter. Sherente oratory was indeed poorly suited to the female voice. I realized that my tribe had irrevocably lost face. I tried to make up for it by telling stories about English wildlife which even Baron Munchausen would have considered exaggerated. But Pia always kept Denmark a neck ahead. I peopled the forests of my country with bears (after all there were quite a few once) and in desperation even fell back on the sabre-toothed tiger. But Pia always capped my wildest flights with Vikingesque tales of wolf packs and hunting deer with horns so wide (elk). The

73

Sherente were more impressed by the size of the elk's antlers, which they could compare with the deer they themselves hunted, than by my fantastic tales of tiger teeth which they justly only half believed. And when my stories got too highly coloured they could always put me in my place by reminding me that after all the chief of my people was a woman.

It was just after sunrise the next morning that Pia shooed away the insistent pigs and marched into the forest. A few moments later she screamed. It was the only time I ever heard her scream. The cry was repeated two or three times and I found myself, still sleepy, running towards the sound, tripping over bushes and roots as I went. I was frightened. It had seemed such a normal day with Wakuke speechifying in the half light and the villagers all hawking and spitting around us. Pia usually went off into the jungle around this time. Indeed we no longer looked on these familiar thickets as jungle at all. We lived here. But those screams suddenly reminded me of the possibilities and even some improbabilities which might lurk among the tired greenery. Sherente too came crashing through the undergrowth to see what had happened.

We met Pia looking white.

'It was a snake,' she said at once, 'a big one. I nearly sat on it.'

The incongruity of the situation was not lost on me, but at the time it did not register.

'Snake,' I shouted in Sherente, 'a big one. Here.'

More men came running with sticks and clubs. Sherente are not frightened of snakes, but they treat them with a healthy respect. That was one of the reasons why they kept their villages so carefully cleared; and the cry of 'Snake!' always brought everybody running to hunt down the reptile and kill it just to be on the safe side. We beat methodically through the undergrowth, making as much noise as we could in the hope that our quarry would flee before us rather than curl up and hide. Women and children came to join in. In a very short time about half the village were threshing through the forest. Pia came with us, albeit a little gingerly. After all it was her snake. I was beginning to think she might have imagined it when there was a shout from one of the beaters. Children jumped as if they were standing on hot stones, others rushed towards the spot. I arrived in time to see a big rattlesnake flowing into the vegetation. It was too quick for its pursuer's club and too stealthy for the rest of us. One moment it was there, rippling across the grass, its big rattle flailing, and the next it had simply vanished. The Indians were persistent though. We poked and pried among the roots and grasses for a good half hour before they finally gave up. Pia had the dubious satisfaction of having her story corroborated and, in retrospect, I could not help seeing the humour of it. Understandably perhaps Pia found herself unable to join in my laughter. The memory of that rattlesnake was going to haunt her every time she slipped into the jungle and she did not find the prospect particularly funny.

Wakuke now joined us. I gave him the stool and some tobacco and squatted with my notebooks close to him. It was a mark of respect which always pleased him, and I was hoping to be able to spend much of the day in conversation with him. But Wakuke was, for once, nervous with us. There was no sparkle in his stories. In fact, there were no stories, for he was unexpectedly laconic. He had something on his mind and we were not long in finding out what it was. He broke into Portuguese in his agitation, for by now he nearly always spoke Sherente with me.

'*Compadre,*' he began, and I knew at once that he was going to ask me for something since Sherente always address Brazilians as Godfather when they are about to beg, 'you are a good man and you know our ways. You have seen how we live here and you have seen that we are a poor people. The *civilisados* take our lands so there is nowhere to hunt, and when we plant our crops their cattle come and spoil them. We are weak people. We cannot fight the *civilisados*, we only fight among ourselves.' He rubbed his fingers together to indicate the friction that existed between Sherente communities. 'The *civilisados* say we are animals because we try to live as our forefathers taught us to live. So now we do not even try. The young men do not study. They forget the ways of our people. They cannot hunt. They run after the tapir like boobies without ever cornering it. They only know how to fornicate and to drink *pinga*. Yes, fornicate! That's all they are good for.'

He was half angry, half amused. Then in deadly earnest he went on, 'We old men, we remember and we try to teach the young ones. If they do not learn to live, then the Sherente will die.'

He managed to infuse the commonplace phrases of his broken Portuguese with a poetic urgency, and I was moved despite myself.

'Now the children go about unnamed, and if we do not name them then the people will forget how these things are done. And we cannot name them because the villages will not come together for the ceremony.'

He paused, hoping that I might say something. The three of us sat in silence for some time. Silences are an important part of Sherente conversation, but Pia and I never quite got accustomed to them. They made us feel slightly uncomfortable.

'If you could give us another bull,' Wakuke said at last, looking down in the dust in his embarrassment, 'we could invite the villagers from Porteiras and hold the name-giving ceremony.'

We did not hesitate. In fact, we had been looking for an opportunity to make some contribution to this ceremony without letting it appear that we were easy dupes for the type of confidence trick which the villagers had played on us with the first steer. Wakuke's eloquent appeal gave us a way of doing this without losing face.

'Wakuke,' I said, 'you are my grandfather, for you have the same name as I have. For you I will do this thing. Invite the men from Porteiras. As soon as I can find anybody who wants to sell an animal . . . and at a reasonable price (I added hastily) . . . I will buy it for you and you can hold this festival.'

Wakuke beamed. No longer was he the wizened, sad little man appealing for a favour which he had good reason to suppose would not be granted. He seemed to sit straighter and the old merry restlessness came over him. He jumped up from the stool and said he would have to tell the others and make the necessary arrangements.

Meanwhile there was less and less to eat in Jacinto's house. I had recently given him more money for food, but he never bought so much as a single consignment of rice with it. Worse still, we were begrudged our share of what food there was in the household. Pia had become expert at wandering round the village and visiting those houses where food had been brought in, just as the Sherente themselves did. She used to bring back whatever she had gleaned and this kept us from starvation, although it did not still our hunger pangs.

Matters came to a head on the day of the first rains. I went out with my shotgun, an ancient blunderbuss with defective sights loaned to me by the Indian Protection Service, in the faint hope of shooting something for the pot. Sizapi came with me to point out a certain species of hawk whose tail feathers, the elders insisted, were essential regalia for the coming ceremonies. The young men could not, for all that, be persuaded to go out after them. We got one solitary bird, but on the way back I shot a howler monkey high up in the horizon of foliage forty feet above our heads. It was, admittedly, a modest bag, but we were bringing back meat and that was the main thing. I had been so engrossed in my quest that I failed to notice the thunder clouds gathering over us. Not so Sizapi.

'We shall get wet,' he said philosophically, 'very wet.'

'Not if we hurry.'

'Yes, we shall get wet. Unless the rain is a long time coming.'

Sizapi clearly belonged to the great tradition of weather prophets. Anyway, we hurried. As we came through the gaping, almost fossilized pores of what was probably a marsh in the wet season it grew dark and a high wind whipped at the trees, breaking off desiccated branches. Lightning flashed overhead, and on the horizon, behind the row of palms which marked the village, we could see an impressive scarlet glow. It was only when we were actually entering the village that I realized this was not one of the storm effects. The huts were etched against flames of cinnamon and purple which made the sky look black by contrast. The whole effect was so stagey that I paused for a second to admire it before it dawned on me that the village plantations must be on fire.

By Jacinto's hut there was chaos and shouting. People were running hither and thither as they always do in the presence of conflagrations, but there was no means of carrying water to the flames in any quantity and the winds which heralded the approaching storm made the fire blaze up as if it were being stoked. We could only hope that the rains would come quickly.

One after another the dry banana trees caught fire, thrusting columns of flame into the sky. Occasionally the wind veered round and the blaze leapfrogged towards the huts. It was only about thirty yards away from the village now and the great gusts of wind could easily have enabled the sparks to hurdle the cleared space separating the huts from the undergrowth.

I met Pia hurrying back towards our hut.

'We had better decide what we want to take out,' she said, 'I don't think we've got much time.'

I grabbed my notebooks and pencils. She took the cameras. On the second trip we took the hammocks. We were just coming back for our trunk when the rain came down like a cataract. Now we had to rush the hammocks and the cameras back under cover again, to say nothing of the notebooks. When it was all done I asked how it had happened.

'That man Jacinto . . . ' she began, and then words failed her as she struggled to describe his various qualities. He had been too lazy to clear the plantations properly or to have them cleared and the undergrowth burned off in the customary fashion before the rains. He had been too dilatory to get on with the burning during the long drought. Today of all days, spurred into activity by the thunder clouds coming over the horizon, he had decided to fire the uncleared growth near the banana plantation.

'I told him that a storm was coming and that the wind would catch the flames,' Pia went on, 'but he just smiled at me as if to say—What do you know about these things? Any fool could see what was bound to happen, but no, not Jacinto. He set fire to the undergrowth less than an hour before the wind rose and at once the fire got out of hand.'

'What's the extent of the damage,' I asked.

'I think he has destroyed the entire banana plantation.'

It was true. Jacinto avoided us that evening. He appeared only to eat a share of the howler monkey which made supper for the entire household. There was no meeting outside his hut either, since it poured with rain all night.

The following morning he approached me as soon as I appeared and said that a certain 'Old Man' in the neighbourhood was prepared to sell a steer. Did I feel like going to see him about it? We would get there when the sun was so high (about mid morning, I calculated). It was a fine morning, glittering after the night's rain, and I quite welcomed the excursion.

'Who will show me the way?' I asked.

'I will come with you,' said Jacinto surprisingly.

'Shall we eat first?'

He looked away from me. 'I thought we would break our fast at the Old Man's place. He has lots of milk and he is sure to give us some.'

I was pretty sure that Jacinto had already broken his fast, probably on the remains of our howler monkey, but I fell in with his plan and set off without saying anything. When we had walked for some time I taxed him about the fire.

'It was the wind,' he explained. 'Oh, what a wind! There was nothing we could do.'

'But anybody could see that there was going to be a storm. Why did you have to burn off your plantation yesterday?'

Jacinto giggled as he always did when he was cornered.

'It burned everything. Everything. Now there are no more bananas. I don't know what we are going to eat.'

'You could try buying some food with the money I gave you.'

'The money was little. You and Tpedi have been here a long time.'

'But you have not bought more than a hundred cruzeiros' worth of food since we have been here and I have given you two thousand cruzeiros.'

For a while Jacinto considered all the possible lines of rejoinder. Finally he slipped into the pose which came most naturally to him. He complained.

'I too am hungry. My family goes hungry because I have to look after my people. Sherente are great ones for asking, and they think that because I am chief they can just ask and ask and I can go on giving and giving. One wants money for seed, another wants money for ammunition. One wants to buy clothes for his wife. Nowadays we cannot go about without clothes like our forefathers did. We have to have dresses for our women and then they ask for combs and perfume and beads, and all these things cost money. And those that have no money, they come to me and say —Kwiro (his Sherente name), you are our chief. You must help us.'

I still did not see why Jacinto should starve us in order to be able to fulfil his chiefly obligations.

'And when they want to celebrate,' he whined on, 'then it is I who have to pay for the fireworks and it is I who have to get a drummer and give him coffee throughout the evening and pay him much money to play for us.'

'Because you are chief, it does not mean to say that you have to buy *pinga* for the whole village, Jacinto,' I said. 'Especially when you do it with the money I gave you for food. Are you not ashamed that guests in your house are not properly fed? What do you think the other Sherente will say when they hear that Tpedi went hungry under your roof?'

'We none of us have enough to eat,' he started on his familiar theme, 'but everybody knows that I am a good man. I know how to receive a guest. I see to it that he is looked after and that he gets proper food like *civilisados* eat. I am generous. I am . . .'

I stopped listening. It was clearly no use arguing with Jacinto.

79

Sometimes I wondered if this was something that Central Brazil did to people. Everybody seemed to have an *idée fixe* to which he returned whenever he got out of his conversational depth.

The Old Man was expecting us. He lived alone in a shack that was if anything rather worse constructed than the huts in the Sherente village. But the interior was very different from the inside of an Indian hut. There was a table, a bench and three cowhide stools. There was dried beef hanging from the rafters, and the pots by the open fire, though hardly spotless, showed evidence of being cleaned occasionally. Best of all, there was a pail of milk on the bench.

There was little conversation. Jacinto and I sat silent while the Old Man made us each a mug of hot milk and flavoured it with a dash of coffee. I found the brew delicious, but it made my stomach rumble so blatantly that I was pleased when we went outside to look at the cattle.

There was a sour-looking bull and a smaller calf waiting for us in a makeshift corral. I examined them with exaggerated care, trying hard to look knowing. I felt sure that the Old Man saw through this pretence, but he was courteous like everybody in the backwoods and gave no indication that he recognized me as a novice where cattle were concerned. Both animals had the right number of legs and, as far as I could see, neither of them suffered from a malignant disease. Both of them were lean by European standards, but not as skinny as some I had seen in Brazil.

'How much do they cost?' I asked.

The old man looked shocked at such directness.

'Which one will you be wanting?' he asked.

'I don't know. That depends on the price.'

This was not true. I knew that I would have to take the bull or else the Sherente would say, as soon as the calf had been eaten, that I had given them the smallest animal I could find and that it hardly provided them with full bellies. Jacinto made no secret of his lack of interest in the calf and the Old Man must have known the position.

We went indoors and sat down to talk it over. We haggled over the smaller animal until I said that if the price was to be so high I would probably get better value for money by buying the bull. Now the Old Man pretended that he did not really want to sell the bull at all. He preferred to keep it for his herd and let us take the calf. The sun was already hanging and the rain clouds gathering again when it was agreed that we should have the bull. The last half-hour of discussion had bored me. I was not much interested in cattle anyway and wanted to get home before the rain came down; but I could not afford to get a reputation in the region as a soft bargainer, so I had fought for the last fifty cruzeiros as if my life depended on it.

The Old Man slipped a halter over our purchase and for the first time in my life I knew what it felt like to own a bull. It ambled along the path in front of us, occasionally straying into the bush where it was not too thick so that we had to head it back again. I got tired of our slow progress. I was already hungry and I didn't want to be wet through as well. So I took the halter and went in front, leaving Jacinto with a switch to bring up the rear. Even so we did not travel very fast. I found myself leaning forward against the rope like a Volga boatman, but if I tugged too hard the bull would stand stock still and pull. Once or twice we had a battle of wills in the middle of the trail. I tugged and cursed while the bull lowered its head and snorted at me, refusing to budge. Then I would shout at Jacinto to give it a whack on the behind, a duty which he performed so gingerly that it made no difference whatever.

We only just beat the rain. The first spattering squalls were falling as we came into the village, preceded by a cavalcade of children. Pia met me lugging my prize along behind me.

'You haven't been pulling it along like that, have you?' she gasped. Pia was reared on a farm, and for a moment I thought I had committed some terrible breach of etiquette.

'What possible difference can it make?' I asked testily.

'But . . . but it's dangerous,' she squeaked.

I looked back at the bull. For the first time I realized why Jacinto had been such a laggard. I put it down to his perennial laziness. For the first time too I really noticed the long curved horns and added them to the sour expression on the animal's face. Up till then I had simply been irritated by its dumb insolence. I dropped the halter and moved away, taking care not to do anything too rapid.

'I didn't know it was,' I said truthfully, missing my opportunity for heroics.

Pia burst into peals of laughter. The Sherente looked at her wonderingly but did not bother to enquire into what had amused her. They were too busy tossing and slaughtering the bull.

'When is the name-giving ceremony to be held?' I asked as soon as I could get anybody to pay attention.

'Tomorrow. The people from Porteiras are coming tomorrow and then we shall have a big festival, just as in the old days.'

We wondered how much meat would be left by the time the proceedings started.

The village was astir well before dawn. No less than three elders were parading up and down, firing hiccoughy speeches at each other across the plaza. Pots of beef stew bubbled in all the huts. We got up early in case we should miss our share.

I had barely finished eating when a man arrived. He made no spectacular entry as Wakuke might have done. He just appeared, and the first time I noticed him he was sitting quietly on the pole in the assembly room, staring at me superciliously. He was strikingly thin, for the Sherente, despite all their moans about having nothing to eat, tend to be corpulent as they grow older.

When I looked at him he greeted me in Portuguese. I replied in Sherente and he looked even more supercilious. I asked him where he had come from and he told me that he had come early from Porteiras. Indeed, he had been one of the speech-makers that morning. It was then that I found out he was Tinkwa and one of the most influential men in Porteiras if not in the tribe.

I had heard quite a lot about him, for whenever I questioned people in the Gurgulho closely about Sherente custom they were likely to tell me to consult Wakuke or Tinkwa. This was peculiarly frustrating, for neither of them lived in the village. There was one old man living in the Gurgulho—the one we had heard on our first night among the Sherente, but nobody ever advised us to consult him. I think he knew this and resented it, for I never managed to draw him out even in trivial conversation. I was therefore particularly anxious to meet Tinkwa; but I had not expected anything like the man who sat with me now. He was shrewd and guarded, even hostile. He had a trick of asking what the men of the Gurgulho had told me about a specific topic, and then ominously failing to comment one way or another when I related it to him. I felt I was being interrogated, and when I remembered that both he and the paramount chief of the Sherente, Dakwapsikwa, came from the same village it did not increase my desire to work in Porteiras.

'People say that you have paid Jacinto much money to teach you about Sherente customs,' he was saying to me.

'That is not true. You yourself must know that Jacinto could not teach me about the customs of your people. Only a *wawen* (elder) can do that and Jacinto is not *wawen*. Why then should I pay him?'

'But you have paid him a lot of money?'

This was a difficult question. If I said 'no', then I would get a reputation for meanness. If I said 'yes', then other Sherente would be jealous and would demand heavy payment whenever they did anything for us.

'I have paid him because Tpedi and I live here in his house and he must buy us food.'

'How much do you pay him?'

I told him, thinking that it would do Jacinto good to have the transaction publicized. Tinkwa was clearly impressed.

'And how much are you paying the people of the Gurgulho to make this festival for you?'

'I am not paying them to make a festival. They feel shame because their children have no names, so I agreed to give a bull that the people might have food to dance and name their children.'

'But it is said that you are paying the people of the Gurgulho to dance for you.'

'Tinkwa,' I said, 'I am tired of you Sherente and your endless gossip and intrigues. You all want presents, but you want them for doing nothing. When somebody is good to me, I give him a present, and at once all the others who never help me, who never give me anything, they come and they say—Why does he get a present when we get nothing? Now you even want me to pay you for naming your own children. Have you come to this? Have you no shame? Do you really think that I would give money to the people of the Gurgulho in order to see them dance? What do I care if they dance or not?'

Tinkwa accepted my tirade without blinking, like a chess player who has pushed his attack too far and watches carefully while his opponent slashes a retaliatory path through his flank. I noticed him studying me calmly as I grew angry. He decided to abandon that particular gambit.

'The Sherente are great gossips,' he admitted, as if that closed the matter. 'It is said that you brought many presents when you came to the Gurgulho.'

I told him exactly what we had brought. I knew he would hear it from the others anyway. He questioned me about what I would bring them in Porteiras and seemed dissatisfied when I told him that it would be exactly the same.

'How about guns,' he said at last. 'Why won't you bring us any guns?'

'They are much too expensive. I have no money to buy a gun myself. How then can I give guns to the Sherente? And what would you do for ammunition anyway, since you have so little money?'

'But they say that you gave guns to the people of the Gurgulho.'

This was really too much. I told Tinkwa, as far as my Sherente would allow me, what I thought of 'them' and by implication what I thought of him too. Then I ostentatiously picked up my notebooks and started writing. But I could not really concentrate. If Tinkwa chose to be angered by my outburst, then I might not only have prejudiced our chances for research in Porteiras but I might also have antagonized a unique informant. Nor could I hide from myself any longer the unpalatable fact that the Sherente were getting on my nerves. Until now I had managed to keep myself emotionally uninvolved and had urged Pia to do the same. When the women reduced her almost to tears with their incessant begging, I advised her to think of herself as a spectator rather than an actor in our every-day dealings with the Indians. When she seethed with indignation over the maltreatment of a child, particularly the little stepdaughter in the next hut, I warned her that we were here to observe and not to interfere, that she must remember we would soon be gone and that if we indulged in the sentimental luxury of trying to 'do good' we would accomplish little with the time and means at our disposal and we would prejudice the detached neutrality on which the success of our research depended. At times this had almost become an issue between us. She did not believe my diagnosis and suspected me of being sanctimonious and perhaps even timid. I had been irritated by her softness which I regarded as the Achilles heel in our intellectual defences. And now I too could grow irritable and be put out, not because I was stirred by injustice but because I was personally piqued. It was disquieting to say the least.

I was still thinking about it when the women arrived from Porteiras. They walked into the village in single file, moving with determination as if they were taking part in a road race. They leaned forward slightly against the weight of carrying baskets suspended from their heads which added to their appearance of competitive zeal. One by one they disappeared inside the huts of their relatives, plonking themselves down inside without ceremony and without greetings.

'Where are the men?' I asked Jacinto.

'They are coming. They are cutting logs in the forest.'

Cutting logs! That meant that they were about to have a log-race. I had read endless descriptions of them dating back to the sixteenth century but I had never actually seen one. I ran into the hut, collected my camera and set off at a trot with Pia following in my wake.

Already the winter marshes were springy underfoot from a few days' rain. There were even puddles here and there in the track which we deliberately splashed through in order to be able to enjoy a few moments of coolness about our ankles before the sun took over again. We waded across a stream which had not existed a week ago and came out on a broad meadow carpeted with coarse grass. From the stunted woods on the other side we could hear the drumming of feet and ululations like a flock of birds.

Soon men came tumbling out of the trees, running, leaping, twisting to see how their companions were faring. Most of them wore nothing but a piece of cloth around their loins, and three of the leaders had long hair down to the middle of their backs which flew out behind them as they came cantering towards us. The front runner looked more like a dervish than a Sherente. His long hair was wavy and its undulations seemed to enmesh the whole of his torso as he ran. All of them glistened with sweat.

In the middle of the throng came the log. It was a heavy length of burity palm borne by a single man who staggered along like a beetle beneath it. As he slowed down another one ran up to him, and the huge log was rolled from shoulder to shoulder. Then the new log-bearer scuttled as best he could while his team mates screamed encouragement or urged him to let them have a go.

They appeared to move very fast. In a matter of moments we were engulfed by the runners. I could recognize nobody, but then I realized that Porteiras were in front so that the first tide of racers were all men whom we had not met. A few yards behind came Kumseran bent double beneath the second log while the men of the Gurgulho urged him to try and pass their rivals before they reached the narrow bottleneck of the forest trail.

Porteiras were the first to reach the village. They flung their log on the ground with an impressive thud and a number of them sat down on it, twitching with exertion. Two old men whom I had never seen before strode up and down, their chests heaving, and sweat running into their eyes, but that did not prevent them from making simultaneous speeches. I gathered as much from the way they strutted about as from what I could understand of their breathless discourse that they were being extremely sarcastic at the expense of the Gurgulho runners. This in itself represented a break with tradition for I was given to understand that the races were utterly uncompetitive. They should have been performed as rites in which everybody was expected to take part and to do his utmost but at which no invidious comparisons were made. There were no answering polemics from the Gurgulho, perhaps because all the men hurried indoors to the beef.

I went in search of the long-haired men, curious to find out who they were and why they wore such a startling coiffure. I found them staying with Sizapi. They turned out to be Kraho Indians from a tribe who lived 150 miles away to the north-east, and they were extremely scornful about their hosts. They could not run; they were useless as log-racers; they did not know how to celebrate a proper festival. Just talk, talk, talk—that was all the Sherente were good for. I was inclined to agree with this verdict, but was more surprised to find that the Sherente present also accepted it quite happily. They smiled good-humouredly like city folk hearing their country cousins extolling the virtues of a rustic life.

Next day they turned the tables neatly on their guests. They challenged them to a race. The challenge was eagerly accepted and it was jestingly suggested that I should join in too. I think both Sherente and Kraho were surprised when I agreed. I laced my tennis shoes on tightly and set out with the runners, wondering if they could see that I already heartily wished I had not been such a fool. My heart sank as we drew further and further away from the village. I was prepared to take on all comers over half a mile. I was prepared to do or die over a mile. After that I was not prepared for anything at all. I was already making excuses for myself in my mind. We had not had enough to eat in the past weeks, I could not be as strong as I had once been . . . I recognized this for a bad sign; and still my companions shuffled briskly ahead. I don't know how far we were from the village when at last they stopped and said that this would be far enough. It was a remarkable understatement as far as I was concerned. The Kraho broke off long blades of grass and bound up their hair. They clearly meant business. I prayed that I would not make a fool of myself by an ignominious showing.

The race started casually and equally casually we loped back towards the village. There were about eight of us, four Sherente, three Kraho and an anthropologist. I forced myself to concentrate on where I planted my feet so as to effect the maximum economy of effort and to take my mind off the contest. The Kraho raced ahead as if they had every intention of disappearing from view. I shall not dwell on the discomforts of the next quarter of an hour. I passed one Kraho walking. He grinned at me, probably amused by my set face. Now we were running through slushy, porous savannah before entering the narrow trail which led into the village. There was a finely-built Sherente running easily beside me.

'Kraho can't run,' he said cheerily.

It dawned on me that we had left the others behind. We entered the village together. The Sherente were jubilant.

'Talk, talk, talk,' they jeered. 'That's all the Kraho are good for. They do not work. They do not plant gardens. All they do is run log races where they come from and yet when they come here they don't know how to run.'

I lay in my hammock concealing my exhaustion and wondering why the Kraho had put up such a poor showing. It was not till later that I remembered the grinning face of the man I had passed. Of course! They had no competitive spirit. They got bored with the race and simply dropped out. They would not have understood the curious motives which had impelled me to run against all my inclinations, let alone outrun them. The Sherente on the other hand had learned the ways of the outside world. They no longer ran for pleasure but only to prove something.

And yet this was still not quite true. Early next morning I set out with all the men for another log race. We gathered in a clearing miles away from the village where the racing logs had already been cut and were now being trimmed. Men with machetes were hollowing the ends of them slightly so that a racer could grip the edge with his fingers. As they did so, others came to try the weight of the logs and make sure that they were roughly equivalent. It was a rough and ready business. By the time we were ready to start the logs still had not been evened out. The young men lifted first the one then the other a few inches exclaiming 'This one's light' and 'This one's heavy', laughing a little at the heaviness of it. Nobody minded.

'*Are kto,*' they said, and two stalwart youths stepped forward. It took four men to lift the logs on to their shoulders, and as soon as they had a hold of them the log-bearers started off down the trail. The lighter log was left behind at the start due to a technical hitch in the loading arrangements, but after a mile or so it had already overtaken the other. I noticed that the men from Porteiras and the men from Gurgulho were all mixed up around either log and there were Kraho in both teams. Not that anybody dawdled. We swept up to Pia who was taking pictures and passed her. She ran alongside for a while but could not keep pace with the logs.

'Wakuke, Wakuke!'

They were shouting at me, trying to get me to carry the log. After the previous day's experience I was prepared to try anything. Besides I did not want it spread around the village that I had refused. I sidled up to the log-bearer feeling rather like a tug approaching a liner. He stopped in deference to my inexperience in the matter of taking a two-hundred-pound log while on the run. In retrospect I remember the discomfort more clearly than the weight. The ribbed bark of the palm trunk tore into my skin and it felt as if chips of bone were being flaked off every nobble in my shoulder. I had noticed that the Sherente carried the log almost behind the shoulder, resting it on the pad of skin at the junction of the neck and back. I tried to imitate them but I did not seem to have any pad there at all, or if I did it clearly was not sufficient. While I was thinking about it, I stumbled and nearly fell. It struck me that a fall with that log might cost me a broken limb and a broken limb on the Gurgulho . . . well, it did not bear thinking about. From then on I looked where I was going and stopped worrying about the pains in my shoulder. I was still jiggling the log about, wondering if I would ever get it comfortable, when a young man who had got his second wind dashed up and presented his shoulder. I felt that honour was satisfied and let him have the log. I trotted into the village among the log-racers and sat down on one of the racing logs which were beginning to clutter up the plaza in order to get my breath back.

There were no speeches. As soon as we had cooled off we dispersed. I felt this was more in keeping with the spirit of the occasion and wondered how long the Sherente would maintain their disinterested involvement in such strenuous ceremonial. There was still no indication of any name-giving. The people from Porteiras had been in the Gurgulho for three whole days and the welcome was beginning to be strained.

Sherente: Men dancing at
the Gurgulho; racing logs
can be seen in the fore-
ground (*above*)

Sherente: Slaughtering a
steer (*right*)

Sherente: Festival at the
Baixa Funda (*above*)

Sherente elders: Pedro the
chief furthest from the camera, next to him Tinkwa,
with Wakuke seated by the
dog (*right*)

Sherente: Wakuke making a speech (*above*)

Leaving the Gurgulho by canoe (*left*)

Just before dusk Tinkwa appeared in the plaza carrying a staff club. He started his speech quietly, and I paid little attention to it, thinking that it consisted either of admonitions which I already knew or ceremonial instructions which I would not be able to understand. But there were murmurs and mutters from the huts as he spoke and then something quite unprecedented happened. Jacinto's senior wife started shrieking from the doorway of her hut. I could not understand half of what she said, but it was obviously highly uncomplimentary to Tinkwa and Porteiras as a whole. I could not understand that Jacinto was not publicly shamed by the fact of his wife's speaking up on behalf of his village. He just sat in the doorway of his hut looking no more than usually ill at ease.

'What is the matter? What does she say?' I asked him.

'The people from Porteiras are angry,' he replied. 'They say that we have not received them well. That we have not fed them properly, that we have not offered them coffee.'

'Coffee! But why should they want coffee?'

'They say that it is shameful to receive guests without offering them coffee. Where can we get coffee from to give them? The people from Porteiras are no good. They only make intrigues and difficulties. We have fed them well. We are generous people. Nobody goes hungry in our village. We know how to receive our guests properly and give them bellies.'

There was much more in the same vein. Pia and I listened incredulously.

'Now they say they will go back to their village,' went on Jacinto. 'Well, I say—Let them go. We do not want them here. We do not want to run about with logs and dance like savages. Here we work like *civilisados*.'

I could not help feeling that the prospects for the Sherente were bleak if two villages could not even come together for a few days without quarrelling over the elementary obligations of hospitality.

That evening, as I expected, a deputation of elders called on me to explain the position. Wakuke was the only one among them who was associated with the Gurgulho. The villagers from Porteiras were restless, they told me. There was not enough food for them in the Gurgulho and they wanted to go home. Yet they were just approaching the climax of the name-giving ceremonies. Tonight the men should have sung until dawn in preparation for the final log race and then the boys would have been named. But the people were hungry and they were in no mood to dance. Could I help them out by providing some food for the ceremonies?

I reminded them that I had given them a bull for just such an eventuality. Why hadn't they kept it and eaten it during the ceremonies instead of guzzling it the moment it arrived?

Why indeed? they retorted. It was the people of the Gurgulho who could not control their impatience. They had eaten all the meat so that they would not have to share it with the villagers from Porteiras. This was not quite true, but so near to the truth that it was not worth arguing about. Then it was up to the people of the Gurgulho to do something about the present crisis, I insisted. The elders looked at each other. Then there will be no name-giving, they said. I suspected they were right but by then I was beyond caring.

Gloom hung over the village. Jacinto was ostentatiously refusing to take any further interest in the business of the name-giving. There was no evening circle before his hut. Only Dbakro and Sizapi sat with him and they conversed in low tones far into the night. The attention of the community was focused on the elders who were directing ceremonial and the elders did not quite know what to do.

Next morning Wakuke and Tinkwa approached me again. They went so far as to admit that the Sherente had behaved badly and abused my trust in the matter of the two steers, but they asked me as a favour to procure at least some food so that the ceremonies could be completed and an open breach between the villages avoided.

'In the old days,' they said, 'each side named the other's children and we were proud during the great ceremonies. Now there is only bickering and shame. Help us, Wakuke, or our children will not know who they are and then we all die.'

I could hardly resist such a plea, and finally I relented in principle, but we were still faced with the problem of how and where we were to obtain more food at such short notice. I would not give any more steers. I was quite prepared to donate a steer to the village of Porteiras when I got there, but I was not going to give another one to the Gurgulho and thus provoke jealousy among the other communities. Eventually we decided that a messenger should be sent over to a neighbouring homesteader to see whether he could provide us with three sacks of manioc flour. The compromise was accepted and the villagers informed. I had the impression that Jacinto's people were not wild with delight.

They realized that the sacks of flour were primarily earmarked for their visitors, so that the new arrangement would simply mean a prolongation of the visit, more ceremonies and the shame of having their guests fed by an outside party.

They made a half-hearted attempt to appropriate some of the food when it arrived late in the afternoon. Jacinto's nephew had been across the Tocantins in the village canoe to buy the flour and he and two of his comrades came staggering up from the river with a sack each. They disappeared into Jacinto's hut with it, but we were waiting for just such a move. I asked Jacinto to have it brought to me in the assembly room. He claimed that he knew nothing about it, so I told him that the sacks had been taken into the back room of his hut and I wanted them brought to me at once. Two of them were reluctantly produced. I asked for the third one as well.

'Is it all for the people of Porteiras?' asked Jacinto.

'Two sacks are for them and the third is for all of us.'

'Shouldn't I keep the third then in my hut?'

'No,' I protested, 'I shall distribute the third myself. This time the people shall know that the food is provided by me.'

Jacinto approached me and whispered confidentially. 'What about the household here? Aren't you going to keep some for us?'

'Yes. I shall keep some for the household and also some for myself so that I can be sure Tpedi will not go hungry.'

The Porteiras distribution took place with a flourish in the centre of the village. The people of the Gurgulho came crossly and shamefacedly to us and complained bitterly over how little they got. But at least the ceremony could now be finished. All the men were summoned into the middle of the village and told of their ceremonial responsibilities. They must neither eat nor drink (not a great sacrifice as they had just done both) but should sing throughout the night. Above all they should not have sexual relations with their wives or even go inside their huts lest they be contaminated by the presence of the women.

Fires were lit in the plaza and we walked round and round them in single file singing the log racing song. I began to have a new respect for Tinkwa. He and Wakuke led the singing, but Tinkwa's fine tenor soared away from us on the high notes. He did not appear to bear me any ill will from our previous conversation, and Dakwapsika, the chief of Porteiras, was affability itself. They swopped stories with Wakuke and were at some pains to educate me in Sherente lore.

'Haven't they told you anything here in the Gurgulho?' they kept repeating. 'Don't they know anything?' and then, scornfully, 'They are not real Sherente here. They do not even know how to make a festival. Who is there in the Gurgulho who can give advice? Only Wakuke *wawen* and even he does not want to live here any more.'

I looked round to see who was going to defend the honour of the Gurgulho and noticed that none of the senior men of that village was present. Jacinto had not attended the singing at all and the rest had slipped away after a few songs. Wakuke and Sizapi were still left, but the former did not disagree with the sentiments expressed and the latter would not have said so if he did.

It got colder and colder. Those of us who were still singing huddled around the fires for warmth, but it was too chilly to sit down for long. After a few moments someone would jump up again and lead off in the next song. Our numbers diminished continually. As soon as the chill set in the younger men seemed to forget the dire consequences of breaking their taboo, the accidents which would surely take place during the following day's log race, the ill luck which would plague all the boys who received their names at the inauspicious ceremony. They crept away to their sleeping mats and the warm presence of their wives. The elders poured scorn on their retreating backs and shouted sarcastic remarks at the darkened huts.

'Behold the strength of our young men! They cannot stay awake even for a single night. How valiant they are when it comes to copulation! They can keep it up for so long. If only they could show the same endurance on the hunt! How strong they are when they go to their women! Yet they pant like children when they carry the log.'

It did no good. By the early hours of the morning there were only five of us left in the middle of the village, and I saw no reason to stay there any longer myself. I made my excuses and went off, my teeth chattering. At least they spared me the bawdy comments.

When I saw the logs we were to carry next morning, I could better understand what all the fuss was about. They stood in the middle of a little glade and in the dim light of morning they looked like anti-aircraft guns. They must have been about twelve feet long. The bark had been left on them to a height of three feet but it had been cleaned and painted with the designs of the two racing groups. The nine-foot-long 'barrels' had been stripped of bark so that they gleamed with an eerie whiteness. They were two-man logs. Indeed it was no harder to race with them than it was with the individual ones, but lifting them on to the first runners and transferring them in full stride were logistic rather than athletic problems. A miscalculation here and the consequences would have been decidedly unpleasant. The elders were grouped around the start, full of jocular foreboding as to the consequences of the young men's incontinence. The race was completed, however, without any broken bones and the two monster pylons were added to the litter of logs which now cluttered up the centre of the village.

Afterwards all the men trooped off into the jungle to fast.

'I can't think why they made such a fuss about getting food,' said Pia. 'They weren't supposed to eat all night last night and they are supposed to fast all day today.'

Our sojourn in the jungle was a survival of the old Sherente tradition which prescribed days of seclusion and fasting for initiated men before important festivals. Now there were no longer initiated and uninitiated and the fasting and seclusion were conveniently reduced to spending all day (until mid afternoon anyway) outside the village. It would have been quite enjoyable had it not been for the insects which plagued us, having taken a new lease of life after the rains. The men sat about, scratching and slapping at themselves and in the intervals working as best they could on their regalia for the ceremony. They made themselves bast sandals and head circlets and entertained each other with stories.

Most of these were about Sun and Moon, a pair of mythological twins about whom there is a whole cycle of Sherente tales. Some of the tales are classics like the elaborate myth of Sun and Moon who go walking together. Everything Sun touches is infused with new life. Trees grow taller, buds open, animals swell up and grow fat when he looks at them. 'That's the way I want the world to be,' he says. Since Sun and Moon can never agree on the way they want the world to be, life is a tussle between good and bad, between creation and destruction and between the warring seasons. It is not surprising that Sun and Moon were also the founders of the moieties, the heroes who established the eternal interaction of opposites within Sherente society itself.

But there were also endless shaggy dog stories involving Sun and Moon and these had no point at all, save that Sun invariably made Moon look foolish and the manner of his doing it reduced the audience to helpless laughter. They gurgled with glee as the climax approached and burst into gales of mirth at Moon's discomfiture. Good story-tellers added new tales in this genre. Wakuke told one about Sun persuading Moon that if he would only allow him to plant his pigs, then they would multiply even faster than in the usual way. Moon trustingly gave him all his pigs (here the audience shouted 'Oh what a fool he is', 'Moon never learns') and Sun feasted on them till he had eaten every one. Then when Moon came and asked to have his pigs back, Sun told him that if he went to a certain spot he would find a forest of pigs' tails. He would just have to pull them up and he would have enough pork to last him for the rest of his days. Moon went off and searched and searched. At last he found the pigs' tails sticking in the ground, but when he pulled them up he found there was nothing on the end of them. Meanwhile Sun went to Moon's house and seduced his wife.

Everyone laughed so much at this anecdote that the elders could hardly get the men to parade into the village in a fitting manner for the naming ceremony. All the men split up into two lines representing the ancient moieties of the tribe. A crier stepped forward from each line and took up his stance in front of it, leaning on a long bow with a big sheaf of gaudily feathered arrows in his hand. Then each man in the line who had a boy to name led him forward and shouted the new name in the crier's ear. The crier shouted it out loud, and all the men in the opposite line bellowed 'He, He! He, He!' in confirmation. That was the theory at any rate. In practice everyone dragged their candidate for naming forward and besieged the crier, who in turn shouted names indiscriminately right and left, all to the accompaniment of perpetual 'He, He'-ing. I found myself grabbed and renamed, but in the din I could not catch what my new name was to be. The whole business was about as intelligible as a crisis in the stock exchange.

Half an hour later it was all over. The criers were ceremonially presented with bowls of manioc flour topped with bits of a slaughtered chicken, or would have been had not a figure completely shrouded in a grass cape suddenly danced into the village and made off with the food. It was explained to me that the mummer represented the Great Anteater and traditionally a number of them appeared at name-givings and stole the food bowls. On this occasion, however, the amount of food was so insignificant that nobody could be bothered to take the part of an Anteater. Only old Waerokran had performed at the last minute.

'What is my new name?' I enquired.

I was told it was Wawenkrurie, meaning literally 'little old man'. I was not flattered.

'What was wrong with Wakuke?' I persisted, slightly aggrieved.

Apparently there were already too many Wakukes, and they wanted to resuscitate the name Wawenkrurie. The last man to hold the name had been tall like me and he had died without passing it on to any of his descendants. So I became officially Wawenkrurie, but it did not really make much difference. After the chaos of the naming ceremony nobody knew anybody else's name any more. Not only had the children been named, but most of the adults had taken the opportunity to change their names too. They refused to answer to their old names at all, saying that that name was finished and now they should be known as such-and-such. When it was brought home to us that we would have to start learning all the men's names over again, we decided that the time had come for us to visit another village. There, at least, we would be learning new names for new people. But before we did that we had one more important engagement by the Gurgulho—our long-awaited stay with Wakuke.

It took us a while to find his new house in the jungle, but when we did we had to admit that it was the most impressive one we had seen among the Sherente. Now it was completed it had an air of spaciousness because its sloping roof did not quite reach the ground on each side, leaving it open all round save at one end where a little thatched room served as Wakuke's private quarters. Not that I ever saw him use it. He slept on a mat in the body of the house together with the rest of us. It was a cozy arrangement, and I had hoped for long conversations with him into the small hours. But Wakuke in his own house was rather different from the Wakuke we had seen visiting the Gurgulho. He was active and restless as only old men know how to be restless. He was forever jumping up to tend the fire or see what his dogs were doing, and as soon as I got him settled he would likely as not fall into a deep sleep punctuated with cataclysmic snores. It was Pia and I who had the long conversations. Just as we were drowsing into sleep Wakuke would start up, wide awake, and stump about the hut rolling cigarettes and coughing until his guts rattled.

I would not have minded this at all had it not been for the fact that time was short. We could not stay with Wakuke indefinitely nor would we have been welcome to do so. And there was another factor which had thrown our calculations out. We knew that there were at least three people living in Wakuke's house. With two of us that made five and we took care to bring enough food to keep five people going for about a week. But when we got to Wakuke's place we found that, coincidentally, six more people had dropped in to stay with him about the time of our arrival. Guests went on arriving all through the afternoon. I took a count that night of the number of people sleeping under the old man's roof. We totalled no less than seventeen.

'What have they all come here for?' I asked Wakuke, being wilfully obtuse.

'They miss you,' he replied. 'They all wanted to be near you and Tpedi.'

'To be near our manioc flour, you mean.'

Wakuke hooted with laughter and then looked at me owlishly.

'Sherente like to eat,' he remarked sententiously, then his waspish temperament got the better of him. 'There is only one thing they enjoy more than eating.' He made a simple and obscene gesture which left us in no doubt as to what that was. We could have guessed anyway, since we already knew Wakuke's views on the modern generation and their propensities.

I felt that I could be outrageous with Wakuke.

'Dogs like these things too,' I said.

He laughed until the tears came into his eyes. Then he sat up and fondled one of his own puppies. There were never less than three dogs in the house at any one time, but it was characteristic of Wakuke that they were never threatened or kicked. The oldest of them was a surly bitch named Intriguer. She bothered nobody but insisted on being left to herself. It was only Wakuke that could stroke her with impunity. She spent the days sleeping and the nights hunting; the others were being trained to follow her example, and the older ones had already caught on. It was their prowess that enabled the old man to be so carefree about entertaining his visitors. When I asked him what we were going to do when all the food was eaten up, he replied cheerfully that the dogs always brought in something. We thought he was joking or bragging or both but it was quite true. Every night they wandered out into the jungle and their slightest bark brought their master running from his hut club in hand to see what they were after. Nobody got much sleep, but there was always a little meat to be had. Once or twice the dogs killed the game themselves and —wonder of wonders—came trotting back with their prize, an armadillo or perhaps a steppe rat. They were accustomed to do this since they always received a portion of every kill that entered the hut, whether they were involved in the hunting of it or not.

On the third day of our visit it poured with rain. Pia brewed some tea and put enough sweetening in it to make it positively emetic. We gave Wakuke a mug of this brew and he was delighted.

'Tell me,' I said, 'how you came to be called Capador.'

Wakuke spluttered (literally). The Sherente are prudish people, despite their occasional bawdy sallies, and *capador* is the Portuguese word for castrator.

'It was when I was working for the da Silvas,' he began. 'In those days I was a young man and I liked to travel. I travelled all over the place. Oh, I went far. Once I went to the cities on the coast and the people all used to turn round in the street and look at me because I had my hair long. Yes, in those days all the Sherente had their hair long like the Kraho, and in the villages the women wore no clothes and the men did not think this was wrong. But then the young men used to be separated from the rest. They had to live in a separate hut until their hair grew long and they were old enough to get married.'

'Did you have to live in such a hut, Wakuke?'

'Yes, I was one of the last. There were only two of us in the hut when it came to my turn and the old men used to tell us that we had to learn to be strong so that we could grow up into good hunters. But the women were terrible. They used to come to the hut by night and put their hands through the thatch. They used to stroke me and ask me to come outside with them.'

'All the women?'

Vanity wrestled with Truth and the match was drawn.

'Most of them,' said Wakuke proudly. 'Of course, I was handsome when I was a young man. They used to say that I had skin as soft as a girl's.'

'But you were going to tell us why you were named Capador.'

'Oh yes. That was when I was working for da Silva in Mato Grosso. Mato Grosso!' He shuddered at the recollection and hid his hands in his armpits. 'That's the coldest place I have ever been in.' He must have seen the incredulous looks on our faces for he added conclusively, 'Mato Grosso is a very cold place. But the da Silvas did not get cold. They used to drink *pinga* like other people drink water, so that they had fire in their veins. I was young then and I did not drink, so the old man used to take out the bottle of *pinga* and say—"If you drank some of this you wouldn't stand around looking so cold, João Pedro." That's what he used to call me. João Pedro da Silva.' He rolled the name on his tongue. 'Then he would take out his revolver and give it to me. He used to say—"When I have finished the bottle, then you can give it back." He didn't trust himself with a gun in his hand and a bottle of *pinga* inside him. But he always kept his other gun, the one on the other side. And nobody tried any tricks on the Old Man, even when he was drinking. He could outshoot them all left-handed even when he was drunk. That's the sort of man he was.

'One day a Bahiano came up river. I think he had killed a man and had to run away in a hurry. But you know what Bahianos are ... '

Bahianos are people from the State of Bahia and they have a reputation throughout Brazil of being charming, fast talkers and untrustworthy.

'Well, he soon had his own horses and used to graze his cattle close to the da Silva's *fazenda*. People began to say that he was making love to the Old Man's daughter. I don't know whether he was or he wasn't. I don't think the Old Man stopped to find out. He just called me one day to go with him and four others from the *fazenda*. We went in the evening so that we would find the Bahiano at home. The Old Man just walked into his house and told him straight out that he had dishonoured his daughter. When the Bahiano started to protest then the Old Man told us to hold him. He fought hard but what could he do against all of us? Then the Old Man took out his hunting knife and ordered us to pull down the Bahiano's trousers. None of us moved. I was very frightened. So the Old Man pulled out his gun and said we had better do as he ordered or he would shoot the lot of us. He would have done it too. They pulled down the Bahiano's trousers and the Old Man castrated him, just like a pig.

'Soon after that I left the da Silvas, but the story got about. Wherever I went people used to say "O João Pedro, the great *capador*". It was only a joke, of course, but I was frightened. If the soldiers got to hear of it they might have tied me up and taken me away. I was just an Indian without any friends to protect me except the da Silvas and I was too frightened to stay there any more. So I went back to my village.'

He made it all sound very matter of fact. It was quite a way from Sherente country to Mato Grosso too.

'How long did it take you to get back to your people?'

'About a year,' Wakuke replied. It wasn't till I came to try and write Wakuke's life story that I discovered he answered all questions of time in the same way. He stayed everywhere for about a year and most of his journeys took him about a year. In this case, though, I think the estimate could not have been far wrong.

That afternoon it rained again, and Wakuke told us about the olden days when the Sherente and the Shavante had lived side by side. It was the first time a Sherente had ever mentioned the Shavante to us, and we were understandably excited at the prospect of gleaning some information about the joint past of the two tribes we had elected to study.

'The Shavante had very powerful seers,' he told us. 'That was before the coming of the white man. Before the Emperor sent his missionaries to us. Then there was no one to stand up to the *sekwa*. Even the chiefs treated them with respect for they could do strange things.

'Once the Shavante and the Sherente came together to make a festival. For days the men sat in the sun. They did not drink and they did not bathe. Their skins turned black and their hair was matted like the lianas in the forest. But still they sat there and by night they danced and the *sekwa* led them.

'Then one morning the Shavante seers got up and took hold of their black arrows. They danced round the village shouting "Hui, hui, hui!", their arrows buzzed in their hands and the men were all afraid. They hid their heads in the sand for fear. Only the Sherente *sekwa* stood up to see what would happen. Out of the sky came a great alligator. When it opened its huge jaws the men could see that it had teeth the size of a jaguar's and each one was as sharp as that of a *piranha* fish. But our seers were not afraid. They caught it by the tail and flung it into the jungle.'

His listeners sucked in their breath with admiration.

'Now the Sherente *sekwa* danced round the village. "Hui, hui, hui!" they grunted. Their arrows buzzed and all the men were afraid. Suddenly a huge boa constrictor appeared out of the sky. It whistled at the Shavante and their seers thought "These Sherente are tough. Their medicine men are very powerful." Then they danced again and called and hundreds of rattlesnakes answered their singing. But the Sherente *sekwa* were not afraid. They seized their black arrows and shouted "Hui, hui, hui!" so that their people could not hear the terrible rattling of the snakes. Then they sang and a cloud of black hornets appeared, buzzing so low over the village that they blotted out the sun.

'Then the Shavante knew that the Sherente were too strong for them and they ran away up into the hills. That is why we call them the *Sakrikwa*, the people who went up.'

'Did the Shavante ever come back?' I asked.

'No, they never came back. They went away towards the setting sun and the Sherente never saw them again.'

'But have the Sherente never visited them since that time?'

'Some men have visited them. It is said that they still speak our language and remember the old ways. Here the men no longer gather in their clubs, for they have forgotten what the clubs are for and the names that our forefathers used to call them by. But it is said that the Shavante still have them and that is why they are so strong.'

Wakuke had one of his rare moments of seriousness.

'Perhaps if we too had fought the whites and refused to accept them . . . perhaps then we would now be strong like the Shavante. We would have our own lands and the white men would fear us and respect us. But we made peace, and look at us now.'

He shook off the thought like a dog shaking off water and stumped suddenly out of the hut. We could hear Intriguer barking somewhere in the undergrowth, and a moment later Wakuke's voice giving a fortissimo rendering of some Sherente cantata as he went out after his dinner.

We left at daybreak the next day. I promised Wakuke that I would send him a pair of trousers. It was the height of every Sherente's ambition to acquire a new suit, or failing that a pair of trousers, for they were inexperienced tailors and trousers were the most difficult things to make. He accompanied us as far as the main trail leading back to the village and for the first time among the Sherente we felt genuine regret as we saw his stocky figure disappear into the bushes again followed by the inevitable dogs.

Next morning we announced our intention of leaving. As soon as the news got around, the village worked itself into a fever of begging and blackmail. Young hopefuls brought us bast necklets and asked us pointedly if we had any more cloth left. Old harridans brought us nothing and simply demanded our blankets because they were old. One resourceful old woman 'gave' us a chicken which, to the general embarrassment, Pia recognized as one which had been given to us a couple of days previously. Finally our bags were packed and our hammocks rolled up, and we sat down to wait for a passing motor-boat to take us up river.

That day of waiting made us think kindly of our involuntary stopover during the flight north. The men lost interest in us and went hunting after one last attempt to borrow my gun in exchange for assurances that there would be no motor-boat before nightfall. The women glared at the remnants of our packed bales and then went silently to their huts. Jacinto's wife asked us what we would do if no motor-boats came in the next few days. I would have replied that it did not bear thinking about, but fortunately I could not put that into Sherente.

At nightfall we were still there. Sleep was impossible for every frogs' chorus sounded like an approaching motor-boat and got us up, grumpily fumbling for torches and lacing on boots. At midnight two young men reported hearing a motor-boat coming 'but still a long way away'. We kept a damp vigil by the river bank but could hear nothing. At dawn they explained that it must have been the drummer at some backwoods party a dozen miles downstream.

Breakfast was a cheerless meal. We had eaten our chicken the previous day, and since we were in that limbo where we were not quite here but not yet gone, nobody offered us any food. Jacinto told us optimistically that he had seen all the motor-boats on the river go upstream after the first rains and that there would be no more coming our way for at least a month. It was always like that every year. Finally our nerve broke and we decided to leave by any available method of locomotion. Walking was out of the question with our baggage, and the village canoe was barely river-worthy. Still, we had no choice. . . .

The Sherente are not and never have been a river-going people. They use the same word for a canoe as they do for a bridge, in other words they look on it as a means of crossing rivers rather than travelling on them. But when we went down to inspect the village boat we were doubtful if it would even cross the Tocantins. It was for one thing entirely submerged, only its prow being held out of the water by the tow rope, like the chin of a bad swimmer in an instructor's hand. The young man who had volunteered to accompany us was not a bit dismayed. He pulled the flat, ungainly thing out on to the bank as if it were an alligator and then pushed it sharply back into the river so that the water slooshed over its sides. By repeating the process twenty or thirty times he finally reduced the water level inside to about six inches. Pia got in and started bailing and I set about patching up the holes. We patched with old rags and bits of burity palm, but it was disheartening work. If you pushed the improvised bung too hard into a crack you were liable to push a hole in the boat.

It was nearly midday when we left. We crept away against the racing stream, the two of us men edging the boat reluctantly forward while Pia bailed for dear life. After two hours' heartbreak we reached a woodman's shack about four miles upstream and there we took refuge from the heat. I pulled our perspiring boat half on to the bank and went off to see if we could borrow another. Fortunately we could for when we put the pride of the River Gurgle back into the water it made a noise like a snapped guitar string and sank out of sight. Luckily its exit was conducted in the same leisurely tempo as all its other activities so we had time to rescue our belongings before it disappeared.

Now our hearts were high. We had a new boat, we had saved our kit and we had every hope that the rest of the journey upstream to the post would be like those summer afternoons we remembered on rivers in England. But we had forgotten the rains. The thunderstorm caught us before we were half-way. I tried to keep Pia's spirits up by reminding her that if we had been in our original canoe the combination of water from below and water from above would assuredly have sunk it. This was small comfort in a situation where we were soaked to the skin and were battling up river in a fog of rain with thunder crashing about our heads and lightning seemingly racing along the surface of the water. Besides, our boatman was a man who had just divorced his wife, and he paddled with that embittered recklessness which psychologists assure us to be the cause of so many road accidents all over the world. Our progress was both erratic and unstable, and it was a relief that we managed to reach the post at all. We were dragging our sodden belongings up the slippery bluff when the first of three motor-boats came chugging past us, all of them going our way.

5

Feuds and factions

Poor Eduardo. On our arrival at the post we reminded him that he had once promised to take us to the Sherente who lived on the banks of the Rio do Sono, some fifty miles east of Tocantinia. The idea did not appeal to him very much now for the rains were beginning in earnest and travelling was something which every right-minded backwoodsman postponed, if he possibly could, until the dry season came round again. But with good grace he set about collecting the necessary horses and harness for us.

We saddled up in the cool of the morning, or what would have been the cool of the morning if the saddling up had not taken so long. Horses are never fed in Central Brazil and consequently nobody would dream of tethering them. They are left to wander about and crop the grass, so that the first thing the traveller has to do each morning is to go out and find his mount. In this case Eduardo had to find three of them, and after that there was still the packhorse to be harnessed and loaded. It was my first experience of coping with a *cangalha* or pack-frame, an instrument of frustration which was to loom large in our lives thereafter. This particular one was not even a good one. Nothing which the Tocantinia post possessed was good of its kind. We covered the horse's back with sacking and then balanced the warped frame with its torn padding on top of it. The girths were frayed, and I discovered that the animal had a belly as if it were about to foal, which would have been surprising for it was an elderly gelding. I considered it a minor triumph when we got the *cangalha* on to it, but slinging the hide bags on to the frame was even more complicated. The thongs which were supposed to support them promptly snapped. When we repaired them and then with a flourish cast the final strap right over the whole contraption, horse and all, we discovered that the load hung at a drunken angle. One bag was too heavy. We would have to unload and repack them both. It was about this time that we finally admitted to ourselves that the simple life which we had hoped would be one of the compensations of field work in Central Brazil was a mirage. We had simply exchanged one set of frustrations for another.

But in Tocantinia we had a stroke of luck. We met an Indian with the tongue-twisting name of Dakukrenkwa, or Dakukre for short, and we promptly enrolled him as our guide. This left Eduardo free to return to the post.

It was a day or two before Dakukre could be persuaded to leave the fleshpots of Tocantinia, a place so tiny that Carolina was a metropolis by comparison. Eventually we pried him loose, and our little convoy set out in the cool before sunrise. Dakukre led the way. Pia followed on a colt and I brought up the rear, leading our solitary pack animal.

We left the thickly forested regions of the Tocantins and travelled through open grasslands dominated by weird sandstone cliffs. The river so dictates the lives of those who live in these parts that it was a strange sensation to turn our backs on it and leave its protective sweep. The people we were to visit faced away from us, clustering around an alien stream. Sherente or not, we felt as if they lived in a foreign country.

I was telling Pia how in Europe we had lost the feeling of journeying—it had all become a matter of seats and reservations and tiresome formalities. This was *real* travelling, with time to jog along and tell each other the Canterbury tales . . . when we came up with Dakukre.

He was sighting carefully along his muzzle-loader at a steppe rat which was sitting in the middle of the trail looking inquisitively at him. But Dakukre had omitted to clean his firing piece for some years and all his frantic muzzle scraping and breech banging was of no avail. Nothing happened. The rat trotted away, stopped, sat up and looked back over its shoulder. Again Dakukre sighted and his blunderbus clicked noisily. The rat lost interest and wandered away. We looked round and discovered that the gelding had lain down in protest under the sacks of salt we were taking to the Sherente and could not get up again.

Dakukre considered our predicament.

'It's too heavy for him,' he said finally. 'He can't carry it.'

'What do we do now?' I asked.

But Dakukre had made his final pronouncement. He stood stock still and waited for inspiration. At that moment I was seized with a spasm of dysentery and retired from the field, leaving Pia as the sole survivor so to speak. It was she who organized the unloading of the gelding and the creation of a secondary load which we strapped on to her colt with everything from my belt to the strap of her shoulder bag. But all our strapping could not make the load stay on what was, after all, a saddle. Every ten minutes or so it began to sag drunkenly to one side. We would hasten to right it. The colt would bolt and come to a stop only when it was dragging our hammocks along the ground and getting them tangled between its legs.

Our progress was thus only a little bit faster than it had been on our famous canoe trip, and it made matters worse when I hung behind to wash my feet in a stream, strode ahead manfully to overtake the party, took the wrong trail and disappeared into the bush. It was two hours later when Dakukre tracked me down, rescuing me from trying to decipher a maze of tracks in the sandy trail I thought I ought to have been following.

Later we lost sight of Dakukre again while ministering to the colt, and were soon after pulled up short by a veritable Piccadilly Circus of interleaved trails, with little idea of which one we should be following. Pia made comments about my abilities as a pathfinder which broke off diplomatic relations between us until we found that silence was a greater strain than conversation. I regarded it as something of a miracle when we actually reached a lone homesteader's shack that night.

It was here that Pia came into her own. While we unloaded the horses, unharnessed them and watered them, she took it upon herself to organize supper. When we returned to the hut, there-fore, we did not have the usual mouthful of manioc flour and then flop into our hammocks. There was coffee steaming in a forest of mugs, a plate piled high with bits of meat, even oranges which she had scrounged from somewhere in Tocantinia. The owner of the hut joined us for the meal. Pia had invited his wife too, but she declined and shepherded her children into the back room like an anxious hen. They too had declined the invitation to join us, which was just as well for their father told us proudly that he had twenty-eight children, twenty-one still alive. I don't think that the entire family lived with him any more, but even half of it might have overburdened our hospitality. Pia went in to give them some sweets and I asked her how many there were.

'I don't know,' she replied, 'I couldn't count them.'

The following day was ominously misty. We got on the trail as quickly as we could, hoping that we might get to our destination before the rain came down, but it was soon obvious that we had lost the race. We heard the thunder crash into the forest close by, or so it seemed, and literally ran the last mile to a deserted house which Dakukre told us was thereabouts.

Only it was not deserted any longer. I was running head down through the first rain squalls to tear away the gate logs when I heard Pia give a gasp. I looked up and saw a man watching us. He wore a pirate beard and I saw the look of astonishment on his face when he saw I was bearded too. But it was not his beard which caught my attention so much as the rest of him. The bandoliers slung just a little too casually from each shoulder across his chest, and full of cartridges, marked him out as a stranger. People in these parts looked as if they were in fancy dress when they took on their bandoliers, and they could rarely afford to slip more than a dozen bullets into them. 'A gunman' was my first thought when I saw his hat with the upturned brim and the well-worn leather of his clothes. But what would a gunman be doing here? There could not possibly be anybody in the neighbourhood whom anybody else would hire a gunman to kill. It was a pretty remote spot. That was it. Remote! He must be an outlaw. Someone on the run and, by the looks of it, someone who was used to being on the run.

There was no alternative but to carry on as if nothing had happened. I pulled open the gate as another man peered out of one of the huts. He wore tatty clothes and was beardless, more or less. But he carried a gun. I realized what an absurd piece of bluff my own gun was. It was for just such emergencies that I carried it, yet I could not hope even to reach it and would not dream of trying conclusions with the characters in front of me. I would, in fact, have been better off without its advertising a spurious belligerence on my behalf.

'*Bom dia*,' I said. 'Do you live here?'

'We are just taking shelter,' answered the pirate.

'Do you mind if we move in too?'

'Come in,' he said with surprising cordiality.

They watched us intently as we unloaded the packhorse and put our baggage indoors out of the rain. I didn't altogether like turning my back on either of them. They asked us no questions as we made ourselves comfortable, and this too was suspicious. Perhaps they were wondering what to do. We certainly were.

'Have you built a fire?' Pia taxed them suddenly.

'Yes, *senhora*,' said the man surprised.

'Then let's have some coffee.'

She got out the pan and went inside the second hut, from which the men had emerged. The traditional drink made us all feel more at ease. I asked the others if they had come far, not so much because I wanted to know as because it would be very suspicious if nobody asked any of the usual questions with which strangers are greeted in the interior. They said they had. They were drovers taking a herd of cattle up north. They planned to take them up to Kraholandia.

'Is that where the Kraho Indians live?' I asked.

'Yes, *senhor*.'

'We shall be going there ourselves later.'

'We shall only be passing through,' added the bearded one hastily. He was evidently the boss. 'We are going further north still.'

After coffee Pia and I slung our hammocks to wait for the end of the shower. I was thankful that my belt, so obviously a money belt, was dangling from the colt's saddle. I could feel its contents against my thigh and only hoped that our disreputable appearance might lead our new acquaintances to think that we had no money to speak of. We watched them as they dozed, occasionally getting up to have a look out and see how the rain was doing.

As soon as it looked like slackening, we continued on our way. I noticed as we went through the gate a second time that the men's horses were unusually good animals and that there were few cattle tracks around the corral.

'That was a very small herd they were taking all that way,' I remarked to Dakukre a mile or so further on.

'That was no herd,' he replied. 'They were rustlers. They have just stolen a few animals and I expect they will sell them as soon as they get a chance to.'

'But where will they get a chance to?'

'As soon as they get somewhere where they are not known.'

'But they are not known here are they?'

'No. That's how I could tell they were rustlers.'

I was still trying to unravel the logic of this statement when we arrived at the village.

We heard the village as we were making our way through the forest. Dakukre stopped and lifted his head like an animal catching a scent. Long, quavering howls came raucous on the breeze and then died away into the tapping of discarded raindrops.

'*Manto der* (somebody's died),' said Dakukre, continuing stolidly on his way.

The groans and cries seemed to fill the whole glade as we turned into the village. Children played unconcernedly between the huts, but there were no adults about. The chief, Suzaure, came slowly to meet us. His eyes were red and rheumy with the extravagance of his mourning.

'I am sad,' he told us immediately. 'My sister is dead. We hoped that Tpedi would come in time to save her, but she died yesterday.'

We mumbled our condolences.

'My people are sad too. Today they put her in the ground and now they weep, for they miss her very much.'

I did not even know that Suzaure had a sister, but I felt some remark was called for so I said in best Sherente fashion that she was a good person and I could understand the general grief. At this point one of the mourners stopped wailing, stood up and came over to us.

'What have you brought us from Tocantinia?' he asked in matter-of-fact tones.

'Salt and sweetening,' I said, indicating our load, 'and many other presents besides.'

I hoped that my vague and sweeping gesture was more convincing than I felt it to be.

The young man examined our bales judiciously. 'You have brought many things,' he said finally.

Pia was nearly as surprised as I was. On the Gurgulho they would have said 'It is little' and left it at that.

Other men came in at this point, some of them with tears and mucus still unwiped on their faces. They wanted to hear the gossip from Tocantinia and above all to get their hands on all that salt and sweetening. They appeared to have switched off their weeping for the night. Suzaure too had his mind on other things.

'It is said that you bought much meat for the people of the Gurgulho,' he began.

I told him the circumstances in which I had purchased the steers and about the bickering over the name-giving festival. Suzaure chuckled until I thought he would choke, while the young men all laughed uproariously.

'Oh, those people on the Gurgulho! They are not real Sherente any more. They don't know how to make a festival. They don't know how to do anything except drink *pinga*.'

Suzaure returned to the matter in hand.

'We do not have much meat here either,' he said.

'Nonsense,' I replied. 'People say that here in the forest there is much game and that you are the only Sherente who eat meat every day.'

'That's so. We have good forests here and they are full of game. But our men have to go out hunting for many days in order to get meat. Now we will not go hunting any more. Our hearts are heavy for my sister and the men will not have the will to hunt. Also,' he continued more practically, 'it will soon be time for the planting of the rice and then we will be too busy.'

'Suzaure,' I told him, 'it is our custom to provide a steer for each Sherente village where we stay, so that the people may know we have love for them in our bellies.' I thought of the blackmail and wrangling that had accompanied the purchase of those steers on the Gurgulho and went on hastily, lest I should lose my composure. 'But remember this: we ask in return that your people should help us and also (Sherente speeches so often ended in an anticlimax) that they should keep some of this salt in order to preserve the meat when it comes.'

'We will keep the salt.' A number of them spoke up. 'It must be a big animal if there is enough to eat and still some left over for salting.'

Pia was a little apprehensive about the pitch-black room in which we were to sling our hammocks. The first time we hung them up and got into them we both landed with a thud on the ground, our combined weight having eased the flexible material right down. I tightened the cords and we tried again. The hammocks now overlapped slightly, so that if one person turned over the other was liable to be suddenly shot out of his sarcophagus. Rice husks chattered above our faces and innumerable Sherente snored around us. Pia shone her torch. Its butter-coloured beam sliced into the air and picked out four little prehistoric monsters a few feet above us.

'Oh no,' said Pia. 'I can't stand those.'

The beam shivered slightly in her hand as I examined the bats. They hung upside down from the beams of the grain shelf, hairy, teguminous. Their eyes were open and caught the light of the torch unblinking.

'They won't harm you,' I said, trying to sound more reassuring than I felt. I too was repelled by the idea of sleeping with these things a bare yard above our faces. 'You know it's a gothic superstition that they like to get into your hair. And yours is too short for that anyway.'

Pia said nothing but I could feel her hammock shake every time those bats fluttered across our faces during the night. I pulled my blanket over my eyes and tried to ignore them. Indeed, I had succeeded when there was a terrific flapping near by. If that was a bat, then it had wings the size of an eagle. I dived frantically for the torch but before I had hold of it a big cock strutted crowing out of the room. A moment later Suzaure started wailing again for his sister.

'Did you sleep?' Pia asked.

'Not much.'

'Nor did I. Let's get up and give a look around.'

The village was not at its best. A dull wet mist obscured the huts and deadened the noises of the forest behind. The little watercourse was muddy and overgrown. We washed dispiritedly and then came back to see if there was any food.

Suzaure produced two eggs and they were boiled for us. We provided coffee. It was a delicious breakfast and did much to revive our spirits. It was only slightly complicated by our trying to share two boiled eggs between three people. Suzaure had immediately accepted our invitation to eat, and in fact it became customary for us to sit down solemnly together every morning and share whatever was going. In the beginning the system worked well. We enjoyed Suzaure's company and he adopted us. But after a few days there were no more eggs in the village and shortly afterwards the supply of tapioca ran out too. After that Suzaure simply had a cup of coffee with us. He never ate any other meal with us, which was very wise of him for after our first week or two we had no other meals to eat.

That was the only time we regretted having taken our professor's advice about supplies. He had done his field work in even more pioneering days when people did not dream of putting up money for anthropological research. He used to ride out with his savings and trade them for a horse or two at the last outpost of civilization before joining his Indians and living entirely as they did. But he had the good fortune to work amongst a gentle and friendly tribe who had provided him copiously with fish and he had got along splendidly. He tended then to be a little impatient with the soft young people who went into the interior nowadays, always worrying about their medicines and their tinned foods.

'Medicines!' I had heard him remark. 'When I was conscripted into the Prussian army I learned two things: to hate the Prussians and to live simply. We had two medicines in those days. Iodine for the outside and castor oil for the inside; and if that didn't do the trick then we used the iodine inside and the castor oil outside.'

Since he was one of the fittest men we knew we had taken his advice. We took rather more than iodine and castor oil along with us but very little food.

In this village at least we hoped that the system might work. At first it did and with full bellies Pia and I were well enough— it was not until later that shortage of food became a serious problem for us. I spent long hours with Suzaure who had appointed himself my teacher. He would don a pith helmet which looked as if it had been minted in an old-style British colony, fix me with his unblinking stare and tell me about his people as if he were reciting a saga. But his eyes bothered me. He had neither eyebrows nor eyelashes, having carefully plucked them out, and this, combined with the mongoloid fold in his upper lids, gave him the look of a benevolent toad. Moreover, he seemed a quiet sort of person to be chief of a Sherente community. I asked him about it one day.

'You are a true *wawen*,' I said, 'for you know how to explain many things. But are there no others here on the Rio do Sono?'

'I am the only one left,' he replied. 'Over the Rio do Sono there were others, but they are all dead now.'

'On the Rio do Sono?' I asked him, surprised. 'But are we not on the Rio do Sono here?'

He explained to me that his village was now known as the Baixa Funda (Deep Hollow). His people had once lived at a village on the very banks of the Rio do Sono but they moved away because they were afraid of the Old Man who spoke harshly to them and used them badly.

'Which Old Man?' I asked, bewildered. There seemed to be a surfeit of Old Men in this part of Brazil.

'Bruwen,' replied Suzaure, and added callously. 'He will soon die. Already his fingers are falling off and his body begins to rot away.'

I had heard rumours that there was leprosy among the Sherente of these parts and now, it seemed, we were living higgledy-piggledy in with the villagers who suffered from it. Worse still, Bruwen was the Sherente, of all Sherente, I most wanted to meet. He had figured prominently in a monograph about the tribe written by a German-Brazilian anthropologist in the 'thirties. If I wanted to talk to him, then, I would have to be quick.

111

'I must see him before he dies,' I said, knowing that this would arouse Suzaure's jealousy. Jealous elders often put themselves out to surpass their colleagues in the matter of giving instruction.

'Bruwen is a bad man,' Suzaure remarked hastily. 'The people fear him. He took their women and when they complained he used magic against them. There was always fighting and quarrelling. So the people came away and built their houses here in the jungle. Here we have good land for planting and every man lives in peace with his neighbour.'

I reflected that Bruwen must certainly be unusual if at the age of sixty or thereabouts, his body rotting with leprosy, he was still a danger to married women and still able to terrorize their husbands. He was clearly no ordinary man that forced a whole village to move away from him; and if he was too quarrelsome for the Sherente, then he must be quarrelsome indeed! I wanted more than ever to visit him, but I was still unhappy at the idea of living with a leper, since these Sherente were so unhygienic in their habits and there would be no hope of privacy or separateness if I went to stay with him. On the other hand, it would be difficult to get him to the Baixa Funda if he was on such bad terms with everybody here. I would just have to wait and see.

I needed to discuss the matter with some influential man other than Suzaure himself. So I hit on the idea of visiting his recently widowed brother-in-law. I was surprised to find him in an impoverished lean-to, a little brushwood hut about three feet high, at the end of the village. He sat there cross-legged and dignified like an Indian fakir, an impression which was heightened by the fact that women were perpetually coming with bowls of food for him.

He told me he was going to leave the village which seemed surprising since he was obviously held in such high esteem where he was. When I said as much he replied gravely that he was everywhere held in high esteem. In fact, all the Sherente villages were competing for his presence. They wanted him in Porteiras. They wanted him in the Funil. They even wanted him to stay here at the Baixa Funda.

'Then why don't you stay here with your people?' I asked.

'These are not my people,' he insisted vehemently.

'But . . . if they ask you to stay . . . and they bring you food?'

'I am an old man,' he explained, 'a true *wawen*. I am respected throughout the tribe. But these are my wife's people. They are Krozake clansmen. I am a Doi clansman. I cannot stay with them for the Krozake hate the Doi and sooner or later they will make mischief against me.'

'Are there none of your own people here?' I asked. 'How is it that they have permitted you, a true *wawen*, to go and live with the Krozake?'

'We were many,' he said, 'and when Bruwen was chief of the village, we were strong. Now Bruwen is sick and the people say bad things about him. Even his own clansmen.' He lowered his voice conspiratorially, although there was nobody else in the hut with us. 'There are even Doi here in Baixa Funda who have joined with the Krozake, who deny their own clan.'

'Who?' I asked amazed.

He ticked the names off on his fingers. There were at least half a dozen of them.

That evening I taxed Suzaure with it.

'When the Indian Agent comes back to the Rio do Sono,' I said, as innocently as I could, 'who will then speak to him on behalf of the people here?'

Suzaure was slow to reply.

'We have no chief,' he said at last. 'I give advice and the people listen, but I am not a chief. If the people want me to speak for them, then I shall speak for them.'

'Then you will be their chief?'

He pretended to misunderstand me.

'That depends on the Agent. If the government make me chief and pay me a salary, then I shall be chief.'

'And if they do not, will you not speak for your people?'

'I shall speak for them.'

'Will the Doi accept that?'

He gave no sign that I had said anything untoward. He stared like a peasant into the dust for a long while. Then he said, 'I will tell you how it is among our people.' He leaned back reflectively as if wondering how to begin, and at last he intoned, 'Listen well! I shall tell a story.' At once all the children within earshot came over to us. Men made themselves comfortable. Women hovered close by.

The story he told was the old, old myth which all the peoples of Central Brazil appear to have in common. He told how a little boy was abandoned on the hunt by his angry brother-in-law and later rescued and befriended by Jaguar. Jaguar took him through the territories of all the birds and beasts until finally they reached his own home. There Jaguar showed him fire and the boy took it back to his own people. From that day forward the Sherente no longer needed to live on rotten wood and raw pulpy foods for they could cook and eat the proper food which they do today.

The tale took a long time in the telling as befitted its importance, for it is really an origin myth, the story of mankind's progression from the animal to the human level, from the raw bounty of nature to the cooked foods of culture. To emphasize that this was the origin of society, the story ends with the Sherente dividing themselves up into their separate clans and choosing distinctive styles of paint for each one.

'At last the oldest and wisest of the *wawen* rose to his feet,' Suzaure continued, 'and he said "I shall paint like this". Then the others were sad for they had already chosen their paint, but his was the most beautiful. There were only a few who painted as he did and they called themselves Krozake.

'From that day the Doi, the fire people, have always been jealous of us Krozake for they know we are the true Sherente.'

There was a murmur of approval as he finished.

'And here in the Baixa Funda, are the Doi still jealous of the Krozake?' I asked.

'They are still jealous,' he nodded.

'And that is why your brother-in-law is leaving the village?'

'He will not leave,' Suzaure answered levelly. 'He is respected here. He has married our woman and soon he will marry another one of our women. Here we are not angry. We live in peace.'

'And if Bruwen summons the Doi clansmen to return to him, will they not go back?'

'They will not go. They are afraid of Bruwen.'

'Then there *are* Doi here who once lived with Bruwen?'

'*Tambwi*,' said Suzaure, meaning 'finished'. I was not clear as to whether the story was finished or the Doi clan had died out. Anyway, Suzaure had clearly finished for the evening. He got up slowly but in a manner which allowed nobody to detain him and walked slowly over to join his wife. A moment later we could hear them cracking nuts in the dark.

Pia and I went back to the bats. We discussed the possibility of leprosy in the village, an issue which now seemed to have been complicated by the clan feud between Doi and Krozake. There did not seem to be anything much we could do to keep clear of leprosy except wash as frequently and as thoroughly as possible; but the clan business was different. I was extremely anxious to find out more about the various Sherente factions, but it seemed as if I was going to have to visit the irascible Bruwen to get the information I wanted. But how could we do that without a guide?

I was still pondering this problem when we were visited by the oldest Sherente I had ever seen. His emaciated body and long staff club formed a capital D as he stood in the doorway, peering blindly about him. He was guided by his son, a rat-faced youth whom I paid to keep track of our old gelding and see that it did not wander too far afield. The old man came up and looked at Pia in some bewilderment, but his son tugged at his elbow.

'He is here. He is here,' he said, piloting him towards me.

The old man gestured vaguely at me.

'He does not speak Portuguese,' said the son; then, raising his voice he shouted, 'The man speaks our language well, father.'

'What?'

'He speaks our language well.'

'I would like to talk with you about the old days,' I said.

'Give him a present. He wants a present,' prompted his son.

'What does he want?'

'He wants a blanket.'

'I have no blanket now. When we have talked much then I will get him a blanket.'

I offered the old man a stick of tobacco. His son took it for him, saying, 'He does not smoke.' A sure sign of conservatism this. I did not know what more to say. Pia went and fetched a handful of sweets and gave them to him. The old man was delighted. Intently he smoothed out the wrappers and hoarded them, sucking sweet after sweet with slushy exuberance.

'He wants a present,' repeated his son. 'He wants cloth.'

'You don't want cloth, do you?' I asked.

'What?'

'Clothes are not good are they?'

'No. They are not good,' said the old man staunchly. Then he suddenly added, 'I go,' and walked firmly off along the trail, sucking busily. His son was obviously disappointed.

'Father is very old,' he said hopefully.

I ignored the remark for it had occurred to me that this independent old man might be the sort of person who would not mind paying a visit to Bruwen. But I did not know how far away Bruwen lived and the old man was pretty frail.

As it happened it was not his frailty but ours which forced us to postpone any visiting we might have planned. After two months food supplies failed almost completely in the Baixa Funda. What food there was in the village was gobbled up as soon as it was brought in or hidden at our approach. But they were not always quick enough when it was Pia who was doing the foraging. She used to swoop like a hawk before the nuts and berries had been shoved into the nearest basket, and these gleanings were an important part of our diet. She brought them back in her pockets and fed me while I took dictation from Suzaure, much to the amusement of whichever Sherente happened to be present. Otherwise we had nothing but a big bowl of rice at midday. We used to divide it into two portions, eating the first one as our main course with a pinch of salt and the second one as dessert with a handful of sugar. Pia always ate it with her fingers, maintaining that the palm of her hand gave it the only flavour that it had. She made fun of my insistence on using a spoon, which she took as further proof of my British refusal to compromise with the jungle. But the rice diet did little for our morale and our tempers became frayed in the persistent rain.

One day she happened to have the rice bowl on the other side of her as we gloomily forced ourselves to eat our stodgy piles. When I had finished the salty portion I asked her if she would pass me the rice. The request infuriated her. I was told that I only had to lean forward in order to be able to reach it myself; and I told her that I was only being polite. She told me that I was being ridiculous; and I told her that there was no need for us to lose our manners just because we were in the middle of the jungle. She left the hut, saying acidly that she had better go and get us some food since I could only be relied on to provide good manners.

I went hunting that afternoon and returned even more irritated, wet through and empty-handed. Pia went out with a group of women to see if she could collect some fruit. She returned with a basket of wild pineapples. The fruit was acid and bitter; still it made a change from rice. But it was apparently too strong for us. Next morning we both had upset stomachs. I began to realize that the circumstances of this research had slowly but surely been sapping our strength. We discovered that we both had worms, so we took the cure for it. Not simultaneously, though, for the potion not only cleans the worms out of your system but removes just about everything else as well, leaving the patient feeling like the skin of a squeezed fruit. After a day in which I barely had the strength to get out of my hammock I felt well enough to look after Pia while she lay in the throes of the drastic cure. Not that 'looking after' entailed doing very much. There was no food to prepare, no medicines to administer. The most I could do was to prepare frequent cups of hot, weak coffee and help her into the jungle when she needed to go.

As soon as we were both fit enough to get about normally again we decided that we would try and regain our strength by walking over and visiting the mysterious little community which people seemed so reluctant to tell us about. Out of courtesy I told Suzaure of our intention and asked him who might like to accompany us. He said he didn't think anybody was going that way just yet. Whereupon we replied cheerfully that in that case we would go alone. We could never have found our way there by ourselves, but I anticipated correctly that a guide would be found, if only to keep an eye on us.

Next day we set out with a young man who had been talked into accompanying us and were surprised to discover that our destination was only an hour's walk away. It was a small group of huts in a clearing which was little more than a hole in the jungle. As soon as we arrived, the women crowded round Pia to see what she had in her rucksack and the men settled down to talk to me.

I looked around for Bruwen but there was nobody who seemed to match the character which other Sherente painted. The men saw me looking around and mistook my motives, for Sherente are very self-conscious about 'living in the jungle'. They launched at once into a defence of their village, assuring me that the land was wondrously fertile where they were and they would never dream of living anywhere else.

'Not even in the Baixa Funda?' I asked.

'Only Suzaure's people live in the Baixa Funda. We are Bruwen's people.'

'Where is Bruwen?' I asked eagerly.

'He lives by the Rio do Sono.'

'But isn't *this* the Rio do Sono?' I asked, almost in despair.

No, it was not. We still had not caught up with Bruwen. These were his outposts so to speak. When I asked them why they did not live with their chief, they replied that they were afraid. It seemed almost a reflex action in these parts for Sherente to add 'We are afraid' whenever they mentioned Bruwen. But the picture was becoming a little clearer. Bruwen and his leprous family were still on the Rio do Sono. The rest of his villagers had left him. Most had gone to the Baixa Funda which was dominated by the Krozake clan. Yet there appeared to be a strong minority of

Bruwen's own Doi who refused to join up with the mild-mannered Suzaure and were too afraid to stay with their own chief.

I still found it hard to see why everybody had been so mysterious about this little intermediate community until a moment later when Pia came up to me and said: 'Some of the lepers are here.'

She pointed across the hut at a woman who had no fingers on her right hand.

'I found out when she shook hands with me. It was a nasty feeling.'

'Well, it's not supposed to be very contagious,' I assured her as cheerfully as I could; but I could not help looking anxiously at the assembled throng and wondering which patches of dirty skin were evidences of infection.

It was impossible to persuade any of Bruwen's people to lead us to him. They all suddenly had urgent commitments elsewhere and assured us that they would come to the Baixa Funda after the rains in order to take us to see the old man. There was nothing to do but return.

Suzaure questioned us closely about our visit.

'The place is full of people who have the same sickness of Bruwen,' I reproached him. 'Why did you not tell me about this?'

He smiled sheepishly.

'If you had known this, you would not have brought Tpedi here. And if she could not come then you would not have come. Then we would have been sad. My house will be big for me when you leave.'

It was a nicely worded compliment such as few other Sherente would have dreamed of paying us. I could not help liking Suzaure even when he deceived us.

'You are right,' I told him. 'It is not good if Tpedi catches this sickness.'

He grinned naughtily.

'We do not fear it,' he said. 'You need have no fear either.'

'You do not fear it because you are ignorant,' I replied, slightly nettled. 'Instead, you fear Bruwen. I do not fear him.'

'You have no cause to fear him,' Suzaure replied evenly.

But still our plans for visiting the terrible old man were frustrated.

I began having fevers that dogged me night and day, and Pia

was exceedingly worried. Then I started to feel nauseated and we suspected that I had jaundice. It was not a pleasant situation. Pia heard that someone was going into Tocantinia and she persuaded him to take a message to the Syrian trader. He was to give a list of my symptoms to the apothecary and ask him to prescribe for me. She hoped then that he would buy the medicines on our behalf and charge them to our account. The medicines and any letters which had arrived for us could be sent back by the same messenger.

The day the man left I got much worse. I was wrapped in every blanket we possessed and all our spare clothes and still felt that deathly cold deep down inside me which is the sure sign of a really high fever. I remember Pia bending over me and saying that a man had come to see me. It seemed a pretty silly thing to tell me, for men were always coming to see me throughout the day. It was the normal routine of life among the Indians, and I felt that she could simply have told him to go away without coming in and informing me first.

'He wants to sell you a cow,' she kept saying.

'Who on earth is it?'

'A Brazilian who wants to sell you a cow.'

'Ask him to come again another day. Tell him I'm too ill to see him now.'

The next thing I knew was that there was a council of some sort being held at my bedside. I could hear Pia saying, 'He's very ill. He can't talk about it now,' and a Brazilian voice mumbling perfunctory condolences. Suzaure was there too, saying something about a very fat bull.

'But he might be dying,' Pia said desperately.

That really woke me up. The Brazilian did not look much concerned. Sickness and death are commonplaces in the interior, and people do not, on the whole, bother to put on a sickroom manner.

'I have to travel,' the man was saying. 'I am going away for a few weeks, so it's your last chance of getting a steer if you want one.'

Suzaure was looking very distressed and I suspected that my fever was only part of the cause for it. I sat up wearily.

'What sort of an animal is it?' I asked.

There was very little haggling. I paid more than the market price for it, but I was in no mood to care.

'Anyway, it will be a good thing to have some decent food for you,' said Pia. But I had already returned to a nightmare world dominated by Bruwen. Bruwen coming to visit me. Bruwen hideously deformed. Bruwen telling me that he was already dead and upbraiding me for not having been to see him. Bruwen. Bruwen.

Pia brought me some meat broth. I could not understand where she had got it from. Apparently two days had gone by since the Brazilian came to see us and the bull had been fetched and slaughtered. Later in the day I tried to eat some bits of meat but they turned my stomach.

That night I sweated so much that I thought somehow I was sleeping out in the rain. When I realized what it was and stumbled out of the hut in a silly attempt to get cool I tripped over what looked like a large rabbit, but it was a bloated toad which bounded away in the moonlight. By then I had fallen flat on my face.

Pia got me back to my hammock, and the following morning I felt a little better. We calculated that it was about time for our messenger to return from Tocantinia, but it was dark before the frantic barking of the village dogs let us know that somebody was coming. The young man walked into the hut and to our intense disappointment handed Pia not a packet of letters but a tiny scrap of paper. On it the Syrian had scrawled:

'I have no letters for your honour, only two parcels. I await your instructions as to what I should do with them.'

I think we both wept with frustration. But he had at least sent me a bottle of some patent medicine which was supposed to be good for jaundice, so the trip had not been entirely wasted from our point of view.

Pia suddenly came to a decision and strode determinedly out of the room. Half an hour later she came back jubilant.

'You know what,' she burst out, 'I went round the village saying that I would pay somebody to go to Tocantinia and get those parcels for us. In the long house at the end the men all asked, "How much will you give?" So I said, on the spur of the moment, fifty cruzeiros (worth about five shillings). Immediately the man who has just this moment come back said, "I'll go, but I'll have to have something to eat first." I gave him his fifty then and there and he said he would leave as soon as the moon rose.'

It rained constantly now which meant that the remains of the slaughtered steer could not be put out to dry. Nor had they been salted properly, for the Indians used up most of the salt I gave them long before I got the animal. The long strips of rotting meat were hung in the only room of Suzaure's hut—the room we occupied. The sweet heavy smell of it made me retch with nausea. I could not even have the satisfaction of eating any of it. In fact, the only thing I could eat at that time was a soup made out of sweet manioc, rather like potato soup. Pia scoured the village for sweet manioc. She begged for it and obtained it occasionally by blackmail. She might even have stolen it for me for all I know. But there was little to be had and I was left once more with plain rice and the smell of bad beef.

The day after our parcels finally arrived, I got up for the first time and was about to try a few steps to see how weak I was when a little girl came running into the hut shouting, 'Sakrbe is dead. Sakrbe is dead.' Nobody seemed unduly concerned. Pia went off to see what had happened, while I came trembling after. Sakrbe was certainly lying motionless on the floor of his hut, but he was not dead—he was dead drunk. The messenger had apparently brought back more than our parcels the previous night.

We went to see what was happening in his hut. Sure enough there was a crowd of Sherente there. They were boisterous but not belligerent. If anything they were a little sheepish when I came tottering in. I had not realized how emaciated I looked.

A bottle of *pinga* was being passed round, but only those who had already paid up were allowed to have a swallow. Our messenger kept a sharp eye on the rest who were just there for fun, but would have taken a pull at the bottle had it come their way by accident. Pia and I had obviously seen the bottle, so they made no attempt to hide their activity but simply pretended that we were not there. They reminded me of schoolboys sharing a pie and all watching keenly to make sure that each one takes a legal bite.

Nor was *pinga* the only thing that our messenger had brought. I watched fascinated as he produced a brand new bible which he must have acquired at the house of the American protestant missionary. Had she made a convert? One who was prepared to take the gospel to his brethren? It would have been an impressive feat. But this was no ordinary prayer meeting. I wondered what he would do with the bible anyway, since I knew he could not read. Solemnly he tore a page out and handed it to the first man in the semicircle. Then he tore out another page and gave it to the second man and so on till every member of the group sat with a page of Genesis in his hands. Then he rummaged in his carrying basket and produced some tobacco which was also distributed. With the solemnity of an important ritual every man rolled himself a cigarette.

I hurried back to tell Pia what I had seen, but she was in no mood for such trifles. She was worried, she burst out as soon as she saw me. Sakrbe was acting violently and it looked as if there might be trouble.

'He's only drunk,' I told her. 'He'll sober up soon enough.'

'It's not that,' she replied. 'He was apparently working himself into a fever of indignation last night, saying he was going to kill Sinan.'

'But Sinan is Bruwen's son,' I protested. 'He does not even live here.'

'All the same, he's staggering about saying, "Let me get at Sinan" or something like that.'

I thought it would be a good thing if someone who commanded some respect were about if Sakrbe chose to run amok. I had heard stories of the way Indians behaved when they had drunk too much of the liquor to which they were not accustomed. Besides, Sakrbe was as strong as an ox, and I did not feel sure that I would be able to calm him down if he came rampaging over to us.

I went to look for Suzaure and found him with most of the village men. I could see that something serious had happened.

'Where is Sakrbe?' I asked.

'He has gone away. We are trying to find him,' they replied.

'But where has he gone to? Why is he so angry with Sinan?'

They did not answer. They sat about, smoking expectantly until Pedro arrived. Pedro was a Brazilian who had thrown in his lot with these Indians. This enabled him to live together with two pretty half Indian girls and to share the village produce in return for his own minimal exertions in the fields. I had noticed that sharing with Pedro tended to be a one-way process, and I mistrusted his motives for going Indian accordingly. Tonight he wore a look of shifty importance and had painted a crude star on his naked chest to emphasize the solemnity of his coming. He was only slightly embarrassed by our presence.

He sat down together with the influential men of the village. It was the first time we had ever seen him do this. They began to question him, and to our astonishment we realized that it was some sort of séance. They wanted Pedro to tell them whether Sinan was really guilty of the things he was accused of. Pedro spun out the proceedings as long as he could. He paced up and down the hut. He sat quite still with his hands over his eyes. He indicated hammily and with all the means suggested by his limited imagination that he was seeing things far beyond the ken of us ordinary mortals. The Sherente watched him and smoked their shaggy cigarettes down until they burned their fingers. Some of them began to drift away. Pedro must have realized that his act was losing its drawing power, for at that instant he announced that he had seen a great deal but it would not come clear to him until he was confronted with the characters in the drama. Suzaure told him that a message had already been sent to Sinan and the proceedings were therefore adjourned until the following day.

Sinan appeared early the next morning. I saw him coming along the path to Suzaure's hut in the middle of a crowd of Sherente. He was gesticulating furiously and his hat was set on the back of his head, giving him the appearance of a small-time political boss rallying his supporters. Sakrbe appeared, the incarnation of a hangover. Everybody was shouting at Sinan. A few minutes later some women entered Suzaure's hut and sat down in the corner with such a flourish that it was quite clear they too were part of the show. It was they who gave me the story.

Sinan had raped a girl, they said, and now he would have to pay. The girl in question sat looking indifferently at the shindy. Nobody was really concerned with her anyway. The points at issue were purely technical. Had she resisted? If not, then she had been seduced not raped, and the indemnity for seduction was considerably less than that for rape. Had she been a virgin in the first place? If not and she had merely been seduced, then there was nothing to fuss about. But the girl's uncles, of whom Sakrbe was one, maintained stoutly that she was a virgin and that Sinan had violated her. They were demanding a terrific indemnity which would have bankrupted Sinan had he been found guilty. Sinan was maintaining equally hotly that the girl had named him simply because he was known to be a man of property. Why didn't she tell the name of the real culprit? Because the real culprit was some shiftless, thriftless youngster who was good for nothing but lying with the girls, and what could her uncles hope to make out of such a one?

We were wondering what sort of a judgement of Solomon Suzaure was going to pronounce when Pedro was brought in. He did not make an entrance as he had done the previous night. In fact we had the distinct impression that he had been dragged to the meeting against his will. The great seer was looking rather careworn and his painted star was faded. He must have realized that he was in a difficult position. He was going to be asked straight out whether Sinan had raped the girl or not and whatever answer he gave was bound to land him in trouble. Sakrbe and the belligerent uncles would not forgive a verdict of innocent and Sinan was even less likely to forget a verdict of guilty which entailed such dire economic consequences. Pedro was likely to have a broken head before nightfall. But if he admitted that he was unable to decide, what would happen to his reputation as a clairvoyant?

For a long time he temporized. He took Sinan's wrists and looked searchingly into his eyes. Sinan stared back brazenly, but I noticed that the sweat came out on his forehead and his hands shook slightly. These signs would not go unnoticed by Pedro, but how would he interpret them?

Then he questioned Sinan.

Had he ever had sexual relations with the girl?

Of course, who hadn't?

There were sniggers all round which the girl accepted indifferently.

Wasn't he the first to have done so?

No. Certainly not. She was not a virgin when he lay with her.

Her mother and her uncles chimed in angrily here. She was. She had been until recently, indeed until Sinan had forced her.

Sinan denied it, and the discussion resolved itself into a minutely detailed disputation concerning the exact condition of the girl when Sinan had had her.

Pedro was in a quandary, but we had to admit that his solution was ingenious.

'She was taken . . .' he said, seeming to pluck his words out of the great void where he saw these things, '. . . against her will.'

The women crowed with satisfaction and the uncles growled their assent. Sinan looked murderous.

'But it was not by this man . . .' pointing to the accused.

Sinan began his 'I told you so's' with obvious relief.

'It was another man . . .'

'Who was it? What do you see?'

They were crowding eagerly about him now and Pedro was making the most of the limelight.

'It was another man . . . let me see . . . I cannot see him clearly . . . oh! It can't be! . . . Yes, it is!'

'Who? Who? Tell us quickly!'

Even the girl looked more interested than she had done at any other stage of the proceedings. Pedro put on his most cringing expression.

'He will kill me if I say his name.'

'No, no, tell us. We shall protect you.'

'I can't. I'm too frightened. I'm a poor *civilisado*. I have no relatives to protect me. He will kill me if I tell you. I know who he is, but I cannot tell you.'

They could not get anything more out of him. They pleaded. They cajoled. They sneered. Pedro was adamant. He knew who it was, but he would not tell them. He was too frightened.

The crowd dispersed slowly, and I could see that Pedro's prestige, such as it was, had suffered a severe slump. But the matter was by no means closed. The credulous were now busy discussing who it could be that Pedro had 'seen' but was too frightened to name. The cynical argued that Sinan had certainly deflowered the girl and that Pedro had invented a cock and bull story because he was too frighteued of him to say so. There was even a small (a very small) minority who felt that Sinan had been wronged by the avaricious uncles and that Pedro was afraid to say so for fear of the girl's relatives.

After the proceedings I took Sinan aside. I told him that I now knew for a fact that he lived with his leprous father and that I wished him to take the old man a message saying that I would soon be fit enough to travel and that we would then come and visit him. Sinan had had a difficult morning, so he did not even attempt to evade the issue. He agreed to take our message and made off. I think he was glad of the excuse to be on his way without any further delay.

We watched field after field being planted with rice. Weeks passed and still there was no word from Bruwen. The only visitor we had was an irate horseman who came galloping into the village one day and reined up like a knight in a sea of snapping dogs. He dismounted with a clink of spurs, fetched the dogs a kick and thrust his way into Suzaure's hut. There he announced that he had come to get his bull. It transpired that he was a local rancher who had gone away for a short trip, only to discover on his return that his cowboy had sold most of his herd and vanished with the proceeds 'in the direction of Mato Grosso'. The indication was rather too vague for pursuit to be feasible and anyway the cow-hand had so many days' start that it would have been impossible to catch up with him. All the disgruntled owner could do was to try and find as many of his animals as he could and see if he could talk the purchasers into accepting that since they had acquired stolen cattle, the transactions were null and void. He was having a hard job. The buyers, by no means wealthy themselves, tended to take the line that they had bought the animals in good faith, had handed over good money for them and would not part with them until that money was returned. It was no concern of theirs that the cash was half-way to Mato Grosso by now in the possession of a dishonest cowboy. That was something between the cowboy and his boss. The boss could rant if he so wished . . . but he would not get his steers back unless he bought them. In our case the situation was simplified by the fact that the bull had already been eaten. We showed him the last remains of its dried meat and I think for a moment he contemplated demanding even that, but he took a closer look and saw the mould that had grown on it and that was enough for him. Pia was patiently scrubbing a little piece of it with soap and a sharp knife, trying to get some small bits which were fit to eat. That was the last straw. He swung on to his horse again and cantered out of the village. The dogs, gathering courage as he left, gave him a savage send-off.

Still there was no word from Bruwen. I went over to my old friend, Suzaure's brother-in-law, to try and find out why there should be no news from the Rio do Sono. To my surprise I found his little shelter deserted. The old man had gone. I made enquiries. Nobody seemed to know where he was. That meant that they would not tell me, for in general everybody knows everybody else's business among the Sherente.

'Why did he go?' I kept asking.

'He was afraid,' somebody said at last.

I took my problem back to Suzaure. Why should a respected man, the brother-in-law of the chief, leave the village so surreptitiously and why should he be afraid?

'Somebody has died,' Suzaure explained. 'It is only a woman; but my people are angry and so he was afraid.'

I begged him to explain, and he told me that news had just reached them of the death of Bruwen's wife. Not the Brazilian girl who was dying of leprosy anyway, but a Krozake woman. She had been perfectly fit until a few days ago and then her stepson Sinan had come back to the house. They had never liked each other and once more they had quarrelled. Old Bruwen took his son's part and they all got very heated. The woman threatened to go back to her own people on the Baixa Funda, and this had incensed her menfolk who said hard things about the Baixa Funda Sherente, that they were ignorant savages, no better than animals living in the jungle.

'How do you know all this?' I asked.

'Because her daughter was very frightened and ran away and told the others what had happened.'

It appeared that when the daughter plucked up enough courage to go back to the house her mother was no longer there. Bruwen said she had died of a fever and showed the girl where she was buried. For some time the news did not leak out, but when it did, nobody believed the story about the fever. The Krozake were convinced that Sinan had killed his stepmother, abetted if not actively aided by the irascible Bruwen. It was the sort of thing that they would expect from those two.

It was a delicate time for Bruwen's clansmen in the Baixa Funda. One or two of them decided that discretion was the better part of valour and went off quietly to join the other Doi. Our chances of visiting Bruwen seemed slimmer than ever. Sinan would certainly not come back to our village until tempers were considerably cooler, and nobody in the village could be persuaded to guide us to the Rio do Sono, right into the lair of the enemy clan.

On December 21st we had to abandon the idea of making this trip. We could not wait indefinitely for the Sherente to patch up their differences. Instead we asked Suzaure to send the trackers out after his colt and began to prepare for our departure. Pia collected all the provisions we had been sent and distributed them amongst the gleeful villagers. By nightfall we were ready to leave, but the trackers returned with the familiar story of having pursued the mare and the flying colt through the jungles and across the steppes without ever having caught either of them.

Next morning they set out again while we waited on our considerably shrunken bales. At midday they still had not come back and Pia began to feel feverish. I thought it was just the nervousness of departure . . . until I took her temperature. It was 103. Sadly I unpacked our hammocks again and settled her once more into the darkened storeroom. Next day she grew worse and was only partly conscious.

I did not know what to do. We had given away the food from our Christmas parcels and so I had nothing to feed her on except rice and tea. She refused the rice, and her temperature remained frighteningly high. I tried scrounging for her and discovered that I had not got anything like her talent for it. I usually returned dispirited and empty-handed. On Christmas Eve she was a little better, but still had a fever.

I spent the whole morning trying to persuade a woman to sell me a chicken. This was harder to do than it sounds, for chickens are really more valuable to the Sherente than money, and by that stage of our stay on the Baixa Funda I had nothing to offer that was more valuable than chickens. Finally I persuaded her to let me have it for five shillings and my only spare shirt on condition that she gave me an egg as well. I took the fowl back to Suzaure's household and told them that I wanted it cooked for supper, when they were all welcome to share it with us. I also stipulated that I wanted a lot of the gravy to make a chicken broth for my wife. That was the best I could do in the way of a Christmas dinner for her.

I kept a sharp eye on the preparation of this repast. As soon as the chicken was swimming in its own juice I went out and took the top of the gravy for Pia. I explained exactly why I wanted it, but that did not prevent the women from giving me some very black looks indeed, and I felt as if I were robbing them. At dusk I boiled the egg and broke it into the clear broth. Then I proudly took Pia her supper and wished her a Merry Christmas. Lying there encased in her hammock she had had plenty of time to think about Christmas Eve in Denmark and to feel really homesick. But she cheered up when she saw the soup. I had managed to keep it a secret from her—my one surprise.

A moment later I wished that I had not. Her disappointment was doubly great when she tasted it and discovered that it was literally a chicken brine. Suzaure's womenfolk had tipped in their reserves of salt in order to make the dish an especial delicacy, and since I had deliberately taken the gravy before they had a chance to water it down to go feed a dozen people I had also acquired a portion which was ten times salted. Pia was disconsolate. And all the time I could hear the men shouting to me to come out and play my guitar.

Finally I left her to cry herself to sleep and went outside into the brilliant moonlight. Halfheartedly I strummed a couple of Christmas carols, hoping that Pia might hear them and that they would cheer her up. But the Sherente complained. They wanted something livelier. So I played them all the bawdiest songs I could think of. They could not, of course, understand the words, but they were delighted by the simple rhythms, and I was able to work off my irritation at this festival which forces the sentimental to pause each year and take stock of themselves.

On Christmas Day Pia's temperature slowly subsided and she demanded that the colt be fetched. She could not bear the thought of remaining in the Baixa Funda a second longer than was strictly necessary. The trackers went out again, but once more the colt eluded them. On Boxing Day they tried once more; but at midday Pia lost patience. She said that colt or no colt she would not spend another night in the Baixa Funda. She would rather walk back to Tocantinia. It was only a bare forty-eight hours since she had been burbling with fever, but there was nothing I could do to dissuade her. Anyway I was anxious to get going myself.

In the early afternoon we loaded up the old gelding and set off on foot. This time, however, we did not have to slink away from the village. Suzaure himself came with us 'to see us well into town' as he put it, and three other men who had errands in Tocantinia decided to accompany us. It was quite a cavalcade. The gelding was leaner and hungrier than ever and had his nose so constantly in the wayside grasses that he frequently got his hooves tangled up in his bridle. But, as I had feared, Pia had overestimated her own vitality. It did not rain so we were roasted by the sun, and after a couple of hours her 'Let's go' mood began to wear off. She had to take frequent rests so that the others and even the horse were soon out of sight ahead of us. At any moment I expected her to fling herself under a shady tree and say she couldn't go a step further. In which case we could look forward to the prospect of a night in the bush without food and possibly a recurrence of her fever in the morning. The very least that could happen would be that even if we kept going at our present rate the gelding would take the opportunity to lie down for a snooze, and we would have to repack his entire load before we could get him to his feet again.

We struggled on in the hope of reaching the backwoodsman's shack where we had spent the night on the way up and taking shelter there. We were still some way away when a premature twilight closed about us and a storm showed every sign of being about to burst. We felt like ants as we made our laborious way forward, Pia leaning on my arm. The hot wind which we were trying to outpace whipped derisively past us. Soon it would drive the first warm spokes of rain into our backs and mould our wet clothes into cold rags around our bodies. I did not want Pia to be out in it when that happened.

We were fortunate. The rain held off till we reached the shack. We had the horse unloaded and our gear under cover before the storm broke. But it was doubly impressive for its delayed coming. Winds tore through the open hut, swinging our hammocks like young boughs. The rain slanted in from the side we had been careful to avoid but every now and again there was a quirk of wind and a sheet of water squirted in from the lee side where we were hanging. We rather enjoyed the feeling of relaxing inside our inner blanket and letting the outer one get wet. It was like sitting by a good fire at home on a winter evening when there is a storm raging outside.

Our reverie was cut short by a fork of lightning which lit up the whole landscape with its naphtha glare; in that split second of light we saw a huge black shape flop out of the thatch on to Pia's hammock.

'It's a spider!' she squealed, diving under the blankets.

I snatched for the torch, couldn't find it at first as I cast around frantically in the dark. Tarantulas are revolting things, and it was as much as I could do to cut them to pieces or stamp on them when I met them by the light of day. This one was just about the largest that I had ever come across; and how was I to deal with it? I could not batter it to death or slash at it with a machete, certainly not shoot at it when it was sitting on top of Pia. Yet I had to do something quickly or Pia might be fatally bitten. Yet if I fiddled halfheartedly with the thing, I might get bitten myself.

Torch in one hand and machete in the other I advanced on it, hoping that I might be able to sweep it on to the ground and deal with it there. All the others were awake now and had scrambled to their feet to see what was going to happen—though they kept at a safe distance. Imagine their delight when I finally caught the animal in the light of my torch and showed it to be an exceedingly battered straw hat which had been dislodged from the thatch by the violence of the storm.

In the morning I had a brain wave. Remembering that Brazilians in the backwoods love having their photographs taken, I asked our host if he would like me to take a picture of him surrounded by his numerous family. The old man was delighted. We delayed our departure until it was light enough to take the picture and on the strength of it he lent us one of his sons and a horse. The boy had to go almost to Tocantinia to collect manioc flour for the family, so it was arranged that Pia should ride the horse until that point while the boy walked with us. He could then ride back with his flour and Pia would only have a few miles of the journey to complete on foot.

The remainder of the journey was comparatively uneventful, save that the gelding lay down every time he could get away from us so that we had continually to chase him to try and prevent him doing so. In the middle of the afternoon the rain came down again and our Sherente companions made an ungallant dash for ε deserted shack a mile or so down the trail. The gelding entered into the spirit of the thing and made his little dash when he was going up a steep rocky hill. He slipped and deposited his cargo in the mud. Pia dismounted and said she felt so awful that she did not know if she could go any further. Suzaure was still with us, but he sat down and said his back was aching, which was understandable since he was carrying the skins of about eight wild pigs for sale in Tocantinia. The gelding, unmindful of the litter of tin plates and torch batteries around his feet, went back to chewing the cud. I remember saying we would be sure to laugh about this later, but I didn't really mean it.

We were wet through when we got to the shack and this was all the harder to bear because it stopped raining soon afterwards. We shared out the last of our hoarded food while we waited—a tin of sardines and a packet of raisins which we had kept for just such an emergency. The Indians were fascinated. They had never seen the inside of a sardine tin before and thought that the fish were very queer-looking specimens indeed. There was five sevenths of a sardine per head and a few raisins. Suzaure liked the raisins so much that he carefully retained the pips and announced his intention of planting enough raisins to choke up the streams near the Baixa Funda. He brushed aside our doubts as to whether raisins would grow at all. Everything grew well in the Baixa Funda he told us. But he was less impressed by the sardines. His fraction of a sardine gave him such a raging thirst that he had to stay behind and drink lengthily at every stream we crossed on the way to Tocantinia.

That night when we parted company with the boy and with Pia's mount, Suzaure was still thirsty. There was no lack of water either. All the trails were flooded by now and as we neared Tocantinia we discovered that it was cut off by a belt of floodwater. We waded gingerly through it, scuffing our feet through the ooze in order to startle any stinging rays which might be nestling there, only to find when we emerged the other side that the gelding had lain down for the umpteenth time and would have to be reloaded.

By the time we reached Tocantinia the romance of Central Brazil was wearing thin, and Pia was very ill indeed.

6

The man without uncles

Morning in Tocantinia, ushered in by Dona Felismena's parrots.
One of them perched on the roof and yodelled at the rising sun,
giving a passable imitation of a man singing through a glass of
gargle.

'Peace, peace, peace!
With peace I'm marching forward.
Peace, peace, peace!
There's peace deep in my heart.'

It was the pious parrot who was singing. His companion hopped
up and down shouting 'Shut your mouth! Shut your mouth!
Shut your mouth!' and giving a realistic rendering of the extremely unladylike laughter of one of the daughters of the house.
Dona Felismena came out, scandalized, with a broomstick. There
was quiet again, but waking was irrevocable.

Suzaure had gone back to his people and we were alone once
more, but I felt more than alone. I had come to rely on the old
man for companionship more than I realized and now Pia too was
away. We had left her to convalesce in Tocantinia while Suzaure
and I paid a diplomatic visit to the Sherente of the Funil. But she
had not convalesced. Instead she had grown steadily worse and
started running such a high fever that Dona Felismena had summoned one of the local businessmen-cum-landowners who had a
medical degree. On New Year's Eve, when Pia took a turn for the
worse, he arrived at the inn to find the villagers already preparing
for the wake.

'They opened the top half of my window,' Pia said (our room
opened on to the street like a horse box), 'and they clustered there
gaping at me, hour after hour.'

Inside the hostelry Dona Felismena was serving coffee for those
who had come to see Pia die. They made way for the doctor as
they would have made way for a priest, and excitedly awaited
those last rituals of medical mumbo jumbo which they were sure
would speed the patient into the next world.

Pia remembered the doctor sitting in the hammock beside her bed and talking kindly to her. Then he got up and turned to the crowd at the window, saying, in the voice of a commissionaire dismissing the end of the queue:

'She's not going to die this time.'

She had gone to sleep turning the phrase over in her mind. *Die* . . . what did he mean by *this time*? Anyway, the crowds went home disappointed and the doctor prescribed for her. But she could not eat ordinary food yet without feeling nauseated. It was a case of anaemia brought on by malnutrition. Obviously there could be no more living with the Sherente for her, not until she recovered her strength at least.

I would have to face the Porteiras Sherente alone. I felt Pia was well out of this particular excursion, but I did not relish the thought of living with them without moral support of any kind. I told myself angrily that I must be getting anaemic too. It was a poor anthropologist who felt nervous of visiting another village, especially when he already knew the people there. I shook off my pessimism and set out for Porteiras with a fresh load of presents and a lone Sherente for my guide. He strode ahead ignoring me and my pack animals.

Every stream we crossed was now in spate and it was I who had to unload the horses, coax them across the torrent and load them up again the other side. The most I could do was to persuade him to support one half of the load while I lifted the other half on, and I felt that, considering it was presents for him and his people I was wrestling with, he was not over-exerting himself.

In the middle of the afternoon we entered a belt of jungle, and I knew we must be nearing the village. This jungle was much in dispute. It looked an untidy and unprepossessing sort of a place, and it was hard to believe that anyone should want it, let alone fight about it; but the extraordinary thing about Central Brazil is that in a region where there are fewer people per square mile than in most parts of the world, men are still prepared to kill each other for land. Nobody had been killed over this jungle, known locally as the Basin . . . not yet. But it was the heart of Tocantinia's Indian problem. It was claimed by at least three ranchers as their territory. The Indians were actually in possession of it, claiming it as part of the lands they had owned since time immemorial. The cattlemen said the Basin was theirs, however. Their cattle grazed there and they claimed that they were making plans to clear and plant it. The Sherente ignored them. The ranchers told the mayor of Tocantinia that the Indians were slaughtering their cattle. The mayor fired off telegrams to Goiania. Goiania was five hundred miles away and nothing happened.

Then the cattlemen 'gave' the timber in the Basin to members of the Baptist congregation in Tocantinia, hoping to embroil them in the dispute. When the churchmen went out to fell it, they were met by armed Sherente and for a while Tocantinia was certain that an Indian attack was imminent. The townspeople were encouraged in this belief by the inflammatory speeches of the mayor, and more telegrams were sent to Goiania saying that peaceful Christians were being massacred by the Indians.

Then, however, the ranchers made a false move. They sent out a posse of adventurers who fortunately met no Indians but came back to town boasting that they had burned a Sherente encampment. This brought in Eduardo. He informed the Protection Service who sent down an inspector from Rio who demanded that the Indians be indemnified, or he would see to it that the government got to hear of what a lawless bunch of cut-throats the inhabitants of Tocantinia were and that would set back local plans for development by a good twenty years. The threat worked and the Indians got their money.

The result of all this was that Eduardo became the best hated man in the region, and the ranchers and the Sherente in the Basin were left with a standing feud which was all the more violent because it had to be concealed.

The Basin certainly looked a dark and violent place as we made our way through it, but the leaky huts scattered among the trees did not give the impression that they housed a band of Indians making a last stand against civilization. The men seemed rather thinner than average and the women even more shrewish. Pedro (Dakwapsikwa), the chief of all the Sherente, met me wearing a pair of white slacks.

When I came to the Basin I was ready to cast him in the role of the villain. Everybody said he was. He was a smooth talker and a hard bargainer. He was clever and I suspected that he was greedy. He was eminently suited to be Mephistopheles, and immediately he began to live up to my expectations of him. He received me correctly but, I thought, coolly. He did not openly sneer at my gifts but he indicated just as forcefully, though more subtly, that he considered them insufficient and he set about extracting promises of more while the first consignment was still being shared out. He managed to convey the impression that it was only his presence which prevented his wives from falling on me like vultures and he hustled me out to my hut as soon as he decently could.

He slept in my hut that night and I assumed that it was simply in order to keep an eye on me, to see what else I had tucked away in my rucksack and to have extra opportunities for dunning me. But in the morning I got a surprise. Pedro rose before it was light and woke me as he went swishing out of our shelter into the long wet grasses of the glade. A moment later I heard him in the distance, voicing a call which soared to a crescendo and then died indecisively on a single minor diminishing note. I leaped out of my hammock, thinking he was going hunting, and peered myopically into the half-dark forest. Pedro was still calling but he appeared to be coming closer. Nobody else was about, though the mumblings and spittings and interchanges between mothers and children proved that the village was certainly waking up. As Pedro came past my hut I was startled to see that he who was usually so elaborately dressed was naked except for a loin cloth. He had a sword club balancing easily on one shoulder, held there by the crook of the arm which was wrapped around it. He disappeared into the misty dawn, still calling. I heard his voice echoing among the trees as he made his way to the far end of the village. Then he started on his way back, only by this time he was speechifying as he walked. It was an impressive performance for a young man. I had never heard anybody but a grey-haired elder attempt to harangue a village, yet Pedro was not only doing it but doing it well.

I went over to his hut and sat down beside his wives who ignored me. The sun was rising when Pedro came in wet with dew. He sat down affably with me and directed that some food should be brought. I don't know if it was simply his unusual costume, but it seemed as if he had temporarily sloughed off his Brazilianized personality. He was no longer Pedro, the schemer, but Dakwapsikwa, the dignitary.

'What were you talking about?' I asked him.

'I was giving advice,' he replied, 'and I was telling the people that they must help you and receive you well.'

This last was true. That much I had understood; but I also knew the Sherente well enough now to realize that this was a formal courtesy which might or might not be meaningless, depending on the dispositions of the people I was dealing with.

Sherente: Hmowen—a friend
who minded our horse in the
Baixa Funda

Sherente: Pedro the chief
and his family (*top left*)

Sherente carrying a racing
log on his shoulder (*left*)

Travelling in the wet season
(*above*)

Suzaure dictating while I
write

'Are there no *wawen* to give advice here?' I asked pointedly. Dakwapsikwa was not offended.

'We have many *wawen*,' he replied. 'Here there are elders who guide us and help us to remember the ways of our fathers. But I too am a chief. I do not yet speak well, but it is my duty to give counsel to my people. I learn from the elders so that when I become old I shall be able to speak as well as they.'

It seemed a remarkable ambition for a man who was reputed to be the most Brazilian of all the Sherente. I doubted whether he was sincere. But surely, if he wished to deceive me, he would have stressed his emancipation as Jacinto had done, not his traditionalism?

Later that day he came to my hut, resplendant in white trousers and a check shirt. He told me that all his people were busy clearing and planting and suggested that I might like to contribute towards their food while they were engaged on such an important task.

'Here we go again,' I thought wearily, but after a little token haggling I offered a sum of money so that he could buy consignments of rice and manioc flour to keep the village going. It was immediately decided that Tinkwa, the very man I most wanted to talk to, should go after the food.

'But why do you send one of the elders,' I asked Pedro in an aside. 'Surely the young men can carry more food back to the village?'

'Nobody here will sell to us Sherente any more,' he explained. 'They say that we are always fighting and that we should not be here in the Basin. They will not sell us food, even if we pay money,' he finished indignantly. 'That is why we have to send somebody like Tinkwa. He is a good talker. Even the *civilisados* respect him. He will bring back food. But how to carry it?'

My horse was the obvious solution as he and I well knew. I offered it to Tinkwa and we spent the rest of the day trying to catch it. It was the first time I regretted having parted with the old gelding who had seen us to the Baixa Funda and back. I had returned him to Eduardo as no longer fit for service and received a healthier horse in his place. The new one was certainly stronger than Old Faithful but far less experienced. Among other things he did not wait ruminatively for somebody to come up and slip a halter over his nose. He did not bolt either. He just walked ahead of his would-be rider, quickening his pace to match that of his pursuer and slowing down again whenever the other slowed down. After I had spent hours walking along behind him making those absurd squeaking noises which the inexperienced always hope will be interpreted as equine endearments I lost patience and tried to catch him with a flying sprint. The horse immediately cantered impressively round in a full circle and went galloping through the village again while the Indians scattered before it. I cursed them for not stopping it and they laughed at me for not catching it. Indignantly I told them to catch it themselves. It was they who wanted to use the horse. They told me they had no time just at present. Tinkwa was philosophical. He said he had much work to do in his gardens anyhow and could easily put off his trip for a few days. Everybody seemed to have forgotten the hunger which had prompted the original request for food.

I need not have worried about not seeing enough of the men in the Basin. None of them would go into their plantations until they had had a good meal to embolden them. If the meal was rice, then it was not usually cooked until nearly midday which meant that the men of the household sat about waiting for their fuel. As often as not it poured with rain in the afternoons and then they would not venture out at all. And then there were saints' days. This was the aspect of Catholicism which the Sherente most clearly understood and most eagerly embraced. On a saint's day nobody worked. Once, in the Baixa Funda, Pia told Suzaure on Santa Lucia's day that this was a big festival in her country. The next morning all the men stayed at home, arguing that as they had worked inadvertently on a saint's day, they now had to make up for it.

What with saints' days and rainy days I heard more gossip in the Basin than I had done in all my previous time with the Sherente. Particularly about the Gurgulho. All Sherente villages were on more or less bad terms with each other, but between the Basin and the River Gurgle there was undeclared war. As soon as people realized that I had been slightly disenchanted with Jacinto and his faction I was deluged with tales of their deceitfulness and outright wickedness.

The Gurgulho was a community without leadership and without conscience, they assured me. Why, Jacinto was not even capable of making a proper speech! He had to pay other people to harangue the village for him. He paid Wakuke until they quarrelled, and Wakuke would not come to the village any more. Now he paid Sizapi who could not speak, and Brurewa (whom Pia and I had nicknamed Caliban), a notorious lecher.

'Lecher?' I asked. This at least surprised me. 'At his age?'

'Oh yes,' they continued eagerly. 'Hadn't you heard? He raped Estiva's daughter.'

I remembered that oppressive afternoon on the Gurgulho; the gargoyle Waerokran perched on top of his hut; the twittering of the deaf mute girl and the obscene frenzy of her mother. Now shaggy old Brurewa had entered the picture. The Gurgulho did not want for grotesques.

'Didn't anybody do anything about it?'

'What could they do? The girl has no uncles here.'

'But the elders, the chief—didn't they do anything?'

The mention of the chief merely aroused their scorn, and they told me exactly what they thought of the Gurgulho and its elders.

I thought it strange that in spite of all this ill-concealed hostility there should nevertheless have been a minor exodus of people going from the Basin to visit Jacinto's community the very next day. Pedro was evasive when I asked him about it.

'They are all going to visit their relatives,' he said.

He did not explain why they all had to visit their relatives simultaneously. The reason was plain to see when the first of them came struggling back a day or two later. There had obviously been a very good party in the Gurgulho. For alcohol even the Basin Sherente were prepared to forget their feuds. Or rather, were prepared to forget old ones. New ones invariably cropped up as soon as they had had a few drinks.

This time had been no exception. Brurewa's son had got so violently drunk that Jacinto ordered him to be tied up. Tinkwa's son had had his own wife enticed away.

'But I thought you had gone with your father to collect food,' I said to the young man.

He looked very woebegone. Perhaps he was wishing he had. Instead he went and stayed with his wife's people who lived on the Gurgulho. When the dancing began he saw that his wife was out there with the men of the Gurgulho.

'What were you doing at this time?' I asked.

'I was drinking *pinga*,' he replied candidly. 'When I saw what was happening I took some more drink to pluck up my courage, then I stepped into the middle and stopped the dancing. I told them that I did not think well of the way they were inciting my woman to behave. She stopped dancing and went back to her parents' hut. Next day they told me that she would not come back with me.'

'You should have brought her back,' remonstrated the more dashing young men in the audience. The husband looked crestfallen.

'What could I do against so many?'

I felt sorry for him. He must have had a shattering hangover and I could understand that he did not feel in the mood to engage the entire Gurgulho single-handed. The alternative was to lose face, and that he had certainly done.

I saw Kumnanse snort with contempt. Short and slender though he was, Kumnanse was not a person that anyone cared to trifle with. In Tocantinia they would not serve him with drink even when he had money. He was one of those volcanic men whom the world approaches diffidently. I had been warned about him. Come to think of it I had been warned about most of the men in the Basin. I had been warned about them so consistently during the preceding months that it was something of a shock to find that I got on quite well with all of them. They were less sanctimonious than those on the Gurgulho, and I realized that the tough act they put on for outsiders was a pose they adopted because they were fighting for their very existence as a community.

That did not, of course, prevent them from fighting each other as well. I was beginning to realize that the politics of factional intrigue were the stuff of life for most Sherente men. That was what kept them so quick-witted and made them such good orators. It also endowed the more forceful Sherente with a self-reliance which was proof even against the corrosive effects of civilization. Kumnanse was a typical example.

'I have no uncles,' he told me one day, and I realized that his whole character was summarized in those two Sherente words. 'My father used to say to me: "You are all right now because people respect me and while I am alive they will respect you. But you have no uncles and when I am gone you will be on your own. Then you will be persecuted." ' After a pause he added. 'He was quite right.' He went on, 'But now people respect me as they respected my father and I shall teach them to respect my son.'

In this he was quite correct. Kumnanse not only had a reputation for being dangerous when he was drunk, but he was also the sort of man that it did not do to cross when he was sober. I knew that there was a long-standing feud between him and two strapping young kinsmen of the chief's, yet so far Kumnanse had apparently had the better of it. To judge by the stories that were told he had knocked one or other of them down on various occasions. Once when they were all full of *pinga* he had flung his enemy's brand new straw hat into the river. That was going too far. The next day a number of men went after him, but when Kumnanse saw them coming he seized his father's sword club and told them to come and get him. The official version of the incident was that Pedro then went in and disarmed both parties, but I suspect that the attackers were glad to have the opportunity not to have to press home the fight. In any case it was common knowledge that certain young men were out to 'get' Kumnanse. They had even shown him the club with which they were going to 'get' him. But Kumnanse remained unimpressed. He remarked drily that he too had a club and that a man who could run away from his attackers (referring to a discreditable episode involving one of his antagonists who had run away from a brawl on the Gurgulho) was not unduly to be feared.

I watched his slight, wiry body as we hunted together and could not help wondering that he was so well able to look out for himself. It must have been from just such a position as mine, a pace behind on a deserted trail, that old Honoro who had something of a reputation for ferociousness himself had felled Kumnanse with a blow on the head. But so far from putting the violent young man out of action the sequel to the story had been that Honoro was now a fugitive who did not dare to return to his village for fear of his life.

It all started, as usual, at a drink and dance party given by some backwoodsmen. Kumnanse had asked if he might dance with Honoro's wife, who happened also to be his own niece. Honoro gave his permission but swore that subsequently Kumnanse had taken the girl outside and let a Brazilian have sexual relations with her in return for money. Kumnanse denied ever having done such a thing and there were no witnesses. Anyway, it was pointed out, if the incident had happened at all then the girl was obviously a willing party to the seduction since she had not cried out for help as she could so easily have done. Honoro's insistence that she was so young and inexperienced, that she did not realize what was happening until too late was regarded by the cynical Sherente as much too far fetched and the matter was dropped. So Honoro took things into his own hands and felled Kumnanse with what should have been a crippling if not fatal blow, when the latter was on his way back to the village. But

Kumnanse sprang up again and clubbed his attacker to the ground. It is said that he would have killed him had not other people intervened. And now Honoro wandered disconsolately up and down the Tocantins, sending occasional messages back to the village saying that he would like to return if only it could be guaranteed that he would be allowed to live in peace and free from fear of attack by Kumnanse. Apparently no one was willing or able to give such a guarantee.

Yet there was loneliness in Kumnanse's strength. Even his wife was not very close to him. It was perhaps for this reason that he deigned to go about with me as much as he did. In fact it was he who was out with me on a protracted hunting trip which took us close to the Indian Agency on the banks of the Tocantins. When I realized how close we were to Eduardo's house I had the idea of borrowing a horse and going into Tocantinia to see how Pia was. I offered to pay him if he would wait for me overnight at the Indian Post and he readily agreed, stipulating only that I should buy him a straw hat in Tocantinia as an additional bonus. Eduardo lent me his own horse and I set off in the gathering dusk, hoping that the animal at least would know the trails to Tocantinia, since I was quite capable of missing them.

I was in a hurry to reach the Porteiras River before it got quite dark, but horses cannot canter in Central Brazil. They either gallop (which was hardly practicable all the way to the crossing) or they move at an excruciating shuffle which jogs the rider up and down until he feels as if he has bruised every bone in his body. And, of course, they walk. We shuffled. Even so the river was just a pale white gleam when I came to the ford. Now in the rainy season it was no longer a ford. The only way of crossing was either to swim it or to holler for a backswoodsman who lived on the far bank. His hut was half a mile away in the trees, so you had to shout loudly, but he usually heard and came down to the water. He would then take rider and harness across by canoe, while the horse swam along behind, a service for which he charged five cruzeiros (about sixpence). This charge was the subject of heated debate up and down the Tocantins. Most people thought it was outrageously high, and many backwoodsmen made detours involving nearly a day's extra travelling time in order to avoid paying the toll.

I shouted in the falsetto way the Indians had taught me. Then I unsaddled my horse and collected all the harness down by the water's edge. For a while I sat there listening while the horse drank, and enjoyed the rustling calm of the forest. Still nobody came. It got so dark that I could barely distinguish the water a few yards away. I knew that the moon was not due to rise before the early hours of the morning. I climbed to the top of the gully and shouted again, even fired a shot in the air for good measure. Nothing happened. He must have heard me, I reflected. If he did not come, then it was because it was too late or too dark. Or perhaps he was away. But he must have a deputy I thought. Surely a son or somebody would look after the ferry in his absence. I was not too hopeful. Arrangements of this sort tend to be casual in Central Brazil.

I began to feel very lonely. Of course I could try and swim the river and I felt that this would be what was expected of explorers, however timid. The more I thought about it, the less I liked the idea. The current was strong and I would have to pull my reluctant horse behind me. I don't like swimming in the dark at the best of times, and I felt that this strange river full of strange things would not be the ideal place to experiment. And of course the harness would get sopping wet. That, I decided, was the crucial factor. I could not swim across. But I did not particularly want to stay either. I had no food and not so much as a box of matches to start a fire. The rustling of the forest did not seem so peaceful any more. I felt it was alive with animals; and I knew that wildcat often came down to drink from the Porteiras. I was morosely scuffling about with my hands, trying to find sticks with which to start the lengthy process of making fire by friction, when I heard clearer and sweeter than any aria the creak and plop of an approaching canoe.

'Good evening,' I called out. 'I was afraid it was too late for you to come out.'

'It is late, yes,' the boatman admitted, 'but I would have come anyway. I always come when I am called. Nobody has been stranded here yet. That's what I told my wife. But she did not want me to go out because of the dogs.'

'The dogs?'

'Yes. There are mad dogs round here. My wife was afraid for me, but I told her that I could not leave a traveller down at the river. What would people say if they heard that the boatman did not come out, mad dogs or no mad dogs?'

'What indeed?'

We loaded the harness into the canoe and pushed off gently with me holding the horse's bridle. The animal hesitated and the canoe slowed down, slewing over in the water, then he plunged in snorting behind us and we lost sight of the bank.

'Is it you, sir, who is working among the Indians?'

'It is, yes.'

'Is it true that many Indians have died already from the dogs?'

'Not as far as I know. What is all this about the dogs?'

'It was a little dog in Tocantinia that started it. It foamed at the mouth like they do, you know, and dashed out into the forest, biting the cattle and the pigs. Then a girl was bitten and she is very ill, but they say that it was another dog that bit her. I think all the dogs here have got the foaming madness and that's why the people are afraid to go out.'

We clambered out the other side, and I noticed for the first time that the boatman wore a machete so long that it brushed the ground when he stood upright. Together we climbed the steep bank.

'I would have come sooner, sir,' he said apologetically, 'but you know how women are. I had to argue with my wife first.'

'I can quite understand that she was nervous,' I said. 'It was very good of you to come out at all on a night like this.'

'I have my responsibilities,' he said, pointing towards the river as if it were his child. 'And I am not very frightened. No dog will get near me in the daytime.' He tapped his machete. 'Only at night it is difficult to see what is coming at you.'

A second later he had disappeared in the direction of his hut and I was taking my horse at a stumbling walk towards Tocantinia. It was difficult to see anything at all, let alone what was coming at us. I wondered if rabid dogs barked. The few stars I could see were like flecks of gold dust scattered by a careless digger, useless for navigation. The barren steppe hummed and crackled with its own stillness. Furiously I jerked upright in the saddle and urged my horse forward and was immediately swept from the saddle by an unseen bough.

Damn the moon, I thought. Never there when it was wanted. Wouldn't it ever rise? I rode for what seemed like hours until I was convinced that I had missed the trail. If the moon never rose, would we two go on riding through a perpetual night into the vast emptiness of Central Brazil, I wondered? The thought made me light-headed, like the idea of a recurring decimal or two mirrors reflecting the same image backwards and forwards forever. Then I saw a wire fence and I knew that I was nearing Tocantinia.

Twenty minutes later I trotted along a wide pebbly track running past the airstrip. Patches of the night were especially dark where the outlying huts were barely perceptible. Not one of them showed any light. Perhaps these people were too poor to use kerosene except in emergencies. Anyway, they were likely to be asleep by now, preparing to start again in the fields before dawn. The first glimmerings came from the mayor's house, faint as a thief's torch. A few other houses shone like glow-worms beside it. I rode through the silent street, avoiding the potholes as best I could, until I came round the corner into the glare of the bar.

Inevitably there was a game of pool in progress. A dozen men were laughing and talking so loudly that they appeared to be shouting while a radio supported them rather fuzzily with music from Goiania. The noise was positively orgiastic after the silence of the steppe. For the first time I understood the magic of the 'saloon' and felt the pull which drew riders into it like moths towards a candle. I reined up and waited, expecting people to stare at me, to ask me where I had come from, what I was doing out at that time of night; but nobody even noticed I was there. I must have stopped just outside the circle of light. To all intents and purposes I was still in the bush. The bottom half of the bar's wooden doors was tight shut. They were obviously taking no chances with mad dogs either. I had the curious sensation of watching human beings close up as if they were fish in an aquarium. At that moment I was sure that I knew how a Sherente felt when he came slouching into the inhospitable streets of Tocantinia.

I rode on. Apart from the bar, there was neither door nor window open in the whole town. It was like arriving in a medieval village. Dona Felismena's place was similarly well shuttered. I took a stick and hammered on the window behind which I hoped Pia was sleeping, nervously preparing a little speech in case some startled or irate traveller should stick his head out instead. There were agitated rustlings inside, followed by a murmuration of diffident voices, but nobody opened up. I thundered on the shutter again. This time it was unmistakably Pia's voice which asked sleepily who was there.

The pleasure of a surprise homecoming is largely concentrated in the first few moments of meeting and incoherent astonishment. All the rest is explanation and anticlimax. In this case I was cheated of the full enjoyment of my arrival by the fact that the combined efforts of all the members of Dona Felismena's household were not sufficient to get the front door open, so that Pia and I had exhausted our conversational exclamation marks before I so much as got inside. Eventually it was found easier to move our boxes and let me directly into Pia's room, since the front door was still defending itself against all comers. The boxes were moved back at once and I realized that I was well and truly inside a fortress.

Dona Felismena made coffee and I was told about the mad dogs. They had been seen at all points of the compass, so that now nobody felt safe wherever they went. They had reportedly infected some of the cattle and lots of pigs and chickens. They were known to have run amok in the Indian villages, so that people were contemplating the prospect of an invasion by hydrophobic Sherente. Tocantinia was shuttered against the plague.

I think everybody was quite relieved that I was not frothing at the mouth myself. They were frankly incredulous when I told them that the rabies had not reached at least two of the Sherente villages. As soon as Pia and I were by ourselves we talked it over and compared notes. It seemed that the whole superstructure of rumour was built on one unpalatable fact. A little girl had been bitten by a rabid dog and brought in to her relatives in Tocantinia in the hope that she could find treatment in the town. The landowner-doctor was away but, as luck would have it, the local chemist did have a series of anti-rabies injections in stock. They had been given to the girl day by day, and after a week or two she appeared to be getting better. They continued the treatment until there were no more injections left, feeling that if some did her good then more would do her even more good. Eventually the girl died.

The dog which was responsible for her death had long since been chased away into the country, but the thought of the slavering sickness it carried with it still haunted the town. It had almost certainly infected some of the animals in the neighbourhood. Some of these had been killed, others fled. As the girl's condition grew worse, people walked in fear of their own chickens. On the night of her death there had been much activity at Dona Felismena's. The son of the house was the local carpenter, and many of the villagers came to drink coffee and watch him hammering the coffin. Then they went in a drove to the house of the dying child. She was spreadeagled in the front room, motionless while life ebbed out of her. Her family kept an eye on her from the back where a bevy of women were making paper flowers to decorate the corpse. Coffee was being served. What with the locked doors, the unaccustomed bright lights burning through the night and the chattering party who had gathered there, Pia said that the house had a raffish air of illicit carnival. She was reminded once more of the middle ages in Europe—of the people who locked themselves up in lonely castles to dance and tell stories until the plague had spent its force. Only here death had somehow got into the house and one felt it had no business to be there, but should have remained decently outside.

In the early hours of the morning the party broke up temporarily while the girl was arranged in her coffin. Then they went back to their coffee. They buried her at dawn. It had been a hasty funeral. The priest was on the other side of the river somewhere, so no mass was said. The procession went hurriedly and well armed to the graveyard and then came streaming back again before the sun had a chance to come out and represent the living. The stragglers hurried into their huts and bolted the doors. Those who had to be out buckled on their machetes. Another day of siege began.

The night I arrived I found the people alert despite the silence and darkness which appeared to shroud the town. Another mad dog was about. Through the cracks of the darkened windows, blurring the haze of the dull candles, men watched. I too took up my station by the window, clutching my German pistol which I knew would never hit a dog in the darkness except by an amazing stroke of luck. But it was expected of me. Anything else would, I am sure, have been regarded as unhelpful, not to say disgraceful.

It was in this slightly ludicrous stance that I heard the contents of our mail as Pia read it to me over my shoulder. When the others abandoned their vigil I slipped into my hammock, but we went on talking; at last I was so over tired and over stimulated by the varied impressions of the day that I could not sleep. My insomnia increased, the more I thought about how early I should leave the next day in order to collect Kumnanse and get back to the Basin. I was just contemplating getting up to leave, sleep or no sleep, when I passed out. Nobody could wake me, apparently, until I came to in the middle of the morning, aroused by a fusilade. Some little mongrel had been shot to pieces by the watchful citizenry.

An hour later I rode out of Tocantinia. The place looked more flyblown than usual in the glare between rainstorms. Some Sherente from the Funil were squatting in the street.

'Wawenkrurie,' they shouted 'give us some money. You never give us anything. When are you coming to visit our village?'

'Soon,' I shouted back, spurring my horse as best I could with the split heels of my tennis shoes. 'I will come soon. Do not fear.' But this time I knew that I lied to them.

Kumnanse and I made our way back to the village in the inevitable downpour. We parted without a word and I crawled into my damp hut. I did not want to get into my hammock for I knew I would only make it all wet. Instead I crouched beside it while a fine spray of rain misted my shelter.

Next morning the first shaft of light pierced my sleep with an insect-laden thrust. I ducked under my blanket to take stock of myself and my work among the Sherente. Perhaps the time was approaching when the most sensible thing to do would be to leave, to go away and work over my notes in peace, to get away from the strain of living in public while I carried on a perpetual battle of wits with my hosts. The truth of the matter was that I was dead tired and that meant that in all likelihood the Sherente were tired of me too. I lay there trying to rationalize my retreat, conscious all the while that the insects and enervation of the Basin had as much to do with it as anything else.

My train of thought was interrupted by a bedlam of shouting and expostulation from a neighbouring hut. I could hear Kumnanse's sulphurous voice and wondered who in the village had had the courage to cross him. Pedro was already rushing to the scene of the squabble, bearing his chiefly club in his hand. By the time I got there he had parted the contestants and I found myself surrounded by a group of women screaming that Kumnanse had tried to beat his wife.

'I did not beat her,' he said to me scornfully. 'I do not beat women. That is why they think they can scream at me and say bad things about me. No man would have the courage to do that.'

'But why is your wife angry then?' I asked.

'She says I have slept with another woman while I was away from the village.'

'But that's absurd,' I remonstrated. 'I was with you the whole time except at the post and there were no women there.'

'Of course it's not true,' said Kumnanse bitterly, 'but you know what women are. You leave them for a little while and they suspect the worst.'

'But how did she get the idea? Nobody even asked me about it and I was travelling with you.'

'She talked to the other women and they told her I went with you so that I could have another woman.'

I was used to Sherente intrigue, or thought I was, but I was startled by the senseless malice of this piece of gossip. It seemed a despicable way to get at Kumnanse.

I decided there and then to leave before the Sherente went sour on me altogether. The preparations were quickly made, for I had so little to take with me. I made Pedro a handsome present to compensate him for any loss he might suffer through our departure, and a few days later Pia and I boarded a motor-boat going downstream. I felt a little guilty over our departure. Of course it was the right thing to do. Our funds had run out. We were in poor health and Pia clearly needed proper medical care. We had spent eight months among the Sherente and I had gathered reasonably good material. Honour ought to have been satisfied. It was my own feelings towards the Sherente that I felt guilty about, though. I had to admit to myself that the Indians had got on my nerves. Even if we had been in perfect health and possessed of endless reserves of cash I would have wanted to leave. It was not until we were on the motor-boat that I could begin to see them in perspective. Who could blame them for their begging and their truculence, for their hypocrisy and their deceit? They were fighting for their lives in a way which people who have never faced the total obliteration of their own society cannot understand. And everything was against them. Still, that did not make them any easier to live with.

As we lay easily in our hammocks and watched the mouth of the Gurgulho with its familiar palms slide past in the gathering dusk, we remembered our misgivings about motor-boats only a few short months ago. Now we felt as if we were travelling in the lap of luxury. We could hardly wait to get back to the metropolitan delights of Carolina.

Shavante

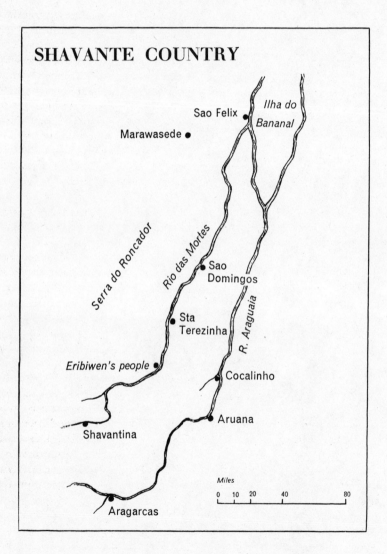

SHAVANTE COUNTRY

Sao Felix

Ilha do Bananal

Marawasede

Serra do Roncador

Rio das Mortes

Sao Domingos

Sta Terezinha

R. Araguaia

Eribiwen's people

Cocalinho

Aruana

Shavantina

Miles

0   10   20   40   80

Aragarcas

7

Paraphernalia

We had been planning so long for our expedition to the Shavante that when the time for it came we were unprepared. Eighteen months after we had sailed light-headedly down the Tocantins we were poised in England ready to return to Central Brazil. But ready is only a manner of speaking. I was by now working for my doctorate at Oxford. We had research funds, generously provided by the Horniman Foundation through the Royal Anthropological Institute. We even had the support of the Brazilian Foreign Ministry which had undertaken to pay my passage and to provide me with a small research fellowship in Brazil. But we also had a baby son and a strong feeling that, in spite of our previous experience, we did not really know what we were letting ourselves in for.

The Shavante were still something of an enigma. The group we planned to visit had been in touch with the Indian Protection Service for four or five years, but nobody was at all sure what the quality of this contact had been like. All we knew for certain was that they were essentially the same band as had annihilated the Indian Service expedition sent into Shavante country a decade previously. At least, we consoled ourselves, they will be comparatively accessible. There was an airstrip near the Indian Agency and a plane was supposed to go up with supplies every week or so. We would, therefore, be spared the journeyings which had proved so exhausting and time-consuming on our previous visit—or so we thought. We could arrange to be flown up to the Indians direct and then . . .

We were not at all sure what we would do then. I did not know whether the Shavante would understand my Sherente, whether they would permit us to live in their community or to travel with them when they moved on. I had nightmares in which they shouted (but always with Sherente voices), 'Don't come here. Keep out. If you come here, we will kill you.' Worse still, I had nightmares in which they all shouted at me and I could not understand a word of what they said. Pia, I think, worried more about the things that might happen to the baby but neither of us communicated these imaginings to the other for we were obliged to present a front of experienced confidence to our friends and relatives. These tended to be either jocular or lugubrious about the responsibility we had towards the child.

This was really the kernel of the problem. It was difficult enough to get ourselves to the Shavante and maintain ourselves there while we did our work, but how were we to manage with a six-months-old baby, even if he should be as much as a year old when we reached the Indians? I was convinced that the Shavante would be intrigued by the presence of Pia and Biorn, our son, and that this might solve at one blow the difficulties of getting to know such reputedly intractable Indians. Even more important, we did not know whether we would do worse by the baby if we deserted him in the second year of his life than if we took him along. We decided eventually to take him. Our insurance against mishap was a battery-operated radio transmitter which we hoped would enable us to summon aid in an emergency and we took with us every medicine we could think of.

We sought medical advice at the Wellcome Historical Medical Museum about tropical diseases. It was a rare pleasure to be able to discuss the situation calmly with the deputy director who gave us a great deal of information and provided us with some drugs as well. His businesslike approach contrasted strongly with that of the various firms we contacted for special items of equipment. Without a single exception they replied either that such equipment could not be supplied (so that we had to improvise it ourselves) or that it would take months to obtain it, by which time we hoped to be deep in the middle of Brazil. It appeared that there were only two stereotypes in the British business world about equipping expeditions. One was the super-technological, along the lines of the expedition to the Antarctic, which was clearly out of our class altogether, and the other dated from African safaris of the nineteenth century. Salesman after salesman fixed us with a glittering romantic eye as we mumbled that we were going to Central Brazil.

'Ah, *yes*, sir. We know all about that. You'll be needing mosquito boots for the little woman I suppose?'

The 'little woman' used to draw herself up to her full height from which, as often as not, she could look down on the salesman and assure him that she could not walk a couple of miles in those things and what she really needed was some form of stout foot-wear that could be slipped on and off easily. This always made the men uneasy. Apparently little women were not supposed to walk about the place once they had equipped themselves with a fly whisk and the house's special brand of mosquito boots. As for slipping them on and off . . . He would pass on hastily to cabin trunks built to keep out the white ants and which would have taken a crane to lift.

'You mean you won't have any bearers, madam?'

It was about this time that the awful truth began to dawn on him. We were just hoboes, not going on a proper expedition at all. Why, we didn't even want a picnic set, and as for all that nonsense about drinking water straight from the rivers . . . If he was fortunate he could leave us here to attend to some young man who was hoping for a spot of shooting where he was going, and we would be left mournfully surveying the fans and fur-belows of a bygone era. He was usually tactful enough not to look in our direction as we tiptoed out.

It was at the elaborately superior West End gunsmith's that Pia's patience finally deserted her. They told us about little guns for big game (perhaps even about big guns for little game—I got mesmerized and do not remember) and were not very respect-fully disbelieving when we assured them that people in the interior of Brazil shot every animal that the country could pro-duce with a ·22 rifle supported by trained dogs. At this point Pia demanded a revolver, the biggest they had. The first one they produced was, she assured them, a peashooter compared with the guns that men carried in Mato Grosso.

'Haven't you got a decent six-shooter,' she demanded aggres-sively.

The produced a short barrelled Webley ·38, the sort they used to tell us in the army was 'accurate up to 75 yards, firing over low cover with both elbows rested' (a situation which I always thought was highly unlikely to occur in action). Pia swung it airily into her hand giving a passable imitation of Annie Oakley and said 'I'll take that one.'

After which we picked up the baby who had been all this time in his carry cot in a corner of the establishment and swept out. She can barely pull the trigger on her revolver and certainly could not hit anything with it at 25 yards, let alone 75 —but it was worth it for the expression on the salesman's face. And as I said to her the first time I heard the thing in target practice, even if it doesn't hit the animal it's bound to frighten it to death.

If the truth be known we too were embarrassed by the title of 'expedition' which had been thrust upon us. But in this respect we were prisoners of the public image of Brazil. Nobody goes to Central Brazil, they make expeditions to it. No wonder the outfitters thought we were impostors. We obviously had not got the right spirit. But in one way the title would, we hoped, prove useful. On our previous visit to the interior we had not had to cross swords with the Brazilian customs, since we had started from São Paulo. This time all our equipment would come under their scrutiny, and they have a reputation throughout Brazil for fickle obstructionism which is equalled only by the reputation of the contraband operators for barefaced cunning. We needed the expedition label to explain to these naturally suspicious functionaries why we needed to import arms, a radio and an expensive tape recorder. We also needed some sort of official document, and we got this through the Brazilian Embassy, done up in green ribbon like a diploma, with only a few days to spare. They informed us that we would also require a clearance permit from the National Council for the Fiscalization of Scientific and Artistic Expeditions to the Interior. The Embassy advised us that our friends could more easily get that for us in Brazil, but the matter proved too complicated for them to arrange in our absence.

So when we landed in Rio nobody would believe that we were an expedition. I produced my sealed parchment from the Embassy. I suggested that they call the Foreign Ministry. The customs were not interested. Our permit was the only possible claim to expeditionhood which they would accept and that we had not got. We asked them to put our baggage to one side, neither in Brazil nor out of it, while we went into the city to sort out the problem. This they were quite prepared to do but the problem proved more intractable than we had feared.

I went to the Foreign Office and discovered that it would take time to get the permit. Nothing had been done about it as yet, and the Council which supervised so many things would not be meeting for some while. By the time I had finished talking to them the customs were closed and we spent our first night in Rio without any baggage. Next day was Saturday and the customs worked only a small shift. They did not get around to our baggage. Pia pleaded with the inspector that we could not spend the weekend in Rio with a baby without any of our luggage.

He said—where nine hundred and ninety-nine Brazilians in a thousand would have been most helpful—that this was no business of his. Finally we did get our personal baggage, but only when the December heat and our importunity and the desire of the officials to get away from it all created an intolerable situation.

Next week we began a merry-go-round of frustration which was to last for months. On Monday I saw a charming old man in a study full of furniture straight out of the colonial period. He was influential on the Fiscalization Council and assured me that they would call a meeting some time before Christmas, that he would do what he could for me then. 'Do what he could for me' sounded ominous—was there any doubt that we would be granted a permit when we had the Foreign Ministry's support and were already in the country waiting to go? On Tuesday I saw the Foreign Ministry, by which time the customs had closed. On Wednesday I saw the customs, but the official went to lunch before he got to our baggage. After lunch he was approached by a professional go-between on behalf of a big American firm, one of whose representatives had just had about £10,000 worth of goods unshipped. Understandably, he elected to go through this consignment first. He got to our items at about 4.00 p.m. and promptly impounded the lot.

At this juncture we decided that there was nothing to be gained by remaining in Rio. Our friends and our work lay in São Paulo, and we could not bear the hotel-room existence in the middle of Rio's enervating hot weather. So we took rooms in São Paulo and waited for our permit to arrive.

Just before Christmas I paid another visit to Rio to try and obtain payment of my Brazilian fellowship and to organize our transport up to the Shavante. The fellowship could not be paid as the Foreign Ministry had closed its annual budget. I was advised that I would have to wait until the following year. I spent the rest of the week trying to get an interview at the Air Ministry, but I only got myself into the position of having the right letter from the right man to the Right Man by the end of the week by which time the Right Man had gone away for the weekend.

On my third visit to Rio I again spent a full week there. I saw the customs who told me that they had been waiting for me to come (though I could not possibly have known it). Now our baggage had been sent to the warehouses and a dossier on it was

in the central customs house. It was impossible to trace the
dossier in the hierarchy, but I was assured that as soon as it
reached the level where a decision could be taken on it then I
would be informed. Meanwhile they got the baggage out for me
to look at again and then casually mentioned that I would not
in any case be allowed to import the arms without a special
permit from the War Ministry. At the War Ministry they were
sceptical about the chances of getting a permit with any speed
but as a special favour, seeing how distraught I was, they presented the documents to the Minister himself and had them
signed later on Friday. They typed out the permit for me on
Saturday leaving me just time to race round to the police with it.
They were to take it and give me in exchange an authorization
to import the firearms. The police too started with the customary
protestations that it was impossible to do anything so late on a
Saturday and ended by taking pity on me and giving me the
permit anyway as soon as I had gone out and bought the right
denominations of stamps to make the document legal.

The minor plot throughout this week was provided by my
negotiations with the Air Ministry. After a day or two they
finally located the place I wanted to go to and informed me that
they would be prepared to authorize our being carried up there
on one of their supply planes, but that in view of the size of the
aircraft they could not permit us to take more than 100 lb of
supplies with us. The rest would be sent up little by little. I
refused, saying that I could not take the risk of jumping into the
middle of Mato Grosso with my wife and baby with so little. It
would not have covered our equipment and the presents for the
Shavante, let alone food, and I did not dare to leave it to the
whim of a lieutenant in Rio to bring us the rest once we were
already in the back of beyond.

On my fourth visit I discovered that the customs were now
demanding hundreds of dollars' worth of duty on the equipment
they had impounded. There was only one thing to do: speak to
the chief inspector himself and ask him if he would intervene.
It took me till Saturday morning to see the great man and he
laughed in my face.

'What do I care if the Foreign Ministry are sponsoring your
work?' he told me. 'The Foreign Ministry should think of these
things when they invite people.'

At the Air Ministry they said they would see what they could
do.

At this juncture I appealed to the British Embassy, thinking that if they made an enquiry on our behalf it might persuade the Brazilians to take a more active interest in our affairs. But our only friend at the Embassy had left and we found it officially unaware that I was in Brazil on a fellowship. They had little interest in the Royal Anthropological Institute or the people to whom it granted research funds. They had apparently heard of Oxford University and wondered if I carried a letter on me to show that I was in fact from that establishment. I assured them that I had requested no identification card from the University on leaving England and suggested that my name and history were now on record in the Brazilian Foreign Office. Finally they went so far as to get one of the attachés to call up the Air Ministry and some weeks later they sent me an official notification in São Paulo that there were indeed Brazilian Air Force planes which carried supplies to the point where we wished to go but that these planes did not normally carry passengers—information which I had known myself before even arriving in Brazil.

The Danish consul-general in São Paulo was more helpful. He gave us a letter to the War Minister. Armed with this I went back for my fifth visit to Rio. Matters now came to a head. The Foreign Ministry decided to cut the Gordian knot by paying the customs what they demanded and thus releasing our baggage. Now it only needed time for the payment to be authorized and the cheque to be issued. When I went to see the Minister of War there was only the matter of transport up to the Shavante outstanding.

The War Ministry in Rio is perhaps the most impressive of all the ministries. It is neither modern nor flashy but it has an air of sober purpose which gives the impression that it is really here that the highest decisions affecting the country are taken. At this time, when the Minister of War was the strong man in the Brazilian government, the impression was probably correct. I waited in line to see his chief of staff and then with some discreet clicking of heels and clanking of ceremonial uniforms a bevy of young officers ushered me into his presence. The General greeted me a little shyly—as indeed I did him for different reasons—but became more loquacious when he discovered that I was capable of carrying on a conversation in Portuguese thus sparing him the necessity of practising his English. He promised to do what he could.

We were now well into the month of March and still stranded

in São Paulo eating up our funds. When I returned for my sixth visit to Rio there was no appreciable alteration in the situation. The War Ministry still had not heard from the Air Ministry, the Foreign Office still had not issued the cheque to cover the customs dues, and the customs, in a final fling, had returned our dossier to the quay in order to have the value of our equipment reassessed in Danish kroner rather than dollars because they had just noticed that we arrived in the country on a Danish vessel.

Next day I presented myself at the War Ministry again and the Minister's chief of staff lost patience.

'We must get this thing resolved one way or the other,' he snapped to one of his officers. 'Take this young man round to the Air Ministry and see what is happening.'

This was quite a new experience. We travelled by staff car and went up in the private lift. The army captain presented the War Minister's compliments and introduced me, and that same afternoon the Air Force worked out a formula. They authorized the transportation of all our heavy baggage on a DC-6 leaving São Paulo for Shavantina in a few days' time. We would then follow in a small Beechcraft which was flying into the interior with Air Force payrolls. The Beechcraft would collect part of our luggage in Shavantina (an hour's flight from our Shavante) and fly it to our destination, then return and take us and the rest of our kit. Our only problem now was to get our baggage out of the customs in time to send it up to the interior.

We managed it with a day to spare. At the last minute it appeared that an unforeseen contretemps might wreck our plans again. The lorry which we had hired in São Paulo to take our baggage to the airport did not arrive, and Pia was in tears at the thought of this final stroke of ill fortune. We could hear the planes coming and going at the airport, and every one of them sounded like an Air Force machine bound for Mato Grosso with our last hopes of getting away on this field trip. Providentially, though, the plane was delayed. The lorry arrived. Our luggage was loaded and our 'expedition' was under way.

8

Shavante at last

'Look,' Pia shouted, unable to contain her excitement, 'look down there!'

The Beechcraft tilted again and we were giants peering right into the middle of a Shavante village. It was just like the old-style settlements we had seen described in countless books by travellers and anthropologists. About twenty big beehive-shaped thatch huts were ranged in a long oval, open at one end. A web of well-used trails glinted like bones as they converged on it. In the centre were two circular patches of cleared ground. It was too good to be true! The meeting places of the two moieties! The plane roared across the river and we could see how broad it was at this point as we doubled back again with the village at our wing tip and began the run in. We sank lower and lower until we could see that the inconspicuous pile on the green carpet below us was thick scrub country. But there was still no sign of an airstrip. We had already lowered our nose when I peered anxiously over the pilot's shoulder and saw the landing strip come at us, ridiculously small like the first casual swathe of a lawn mower whose owner has gone back into the house. The plane bumped to a stop and I jumped out, more excited than I can remember. Shavante were already converging on us.

They were mostly boys by the looks of it, wearing nothing but cylindrical plugs in their ears and a minute conical sheath on their genital organs. They gathered, laughing and joking, around the plane while all our kit was taken out. I realized with a sick feeling in the pit of my stomach that I could not understand what they said. Some of the men were arriving now—presumably those who would not be bothered to rush too hastily to see an aeroplane. I did not have the courage to try my Sherente on them right there. Instead I pretended to ignore them while I ensured that every item of our baggage came out and helped Pia and the baby out of the plane. There was no sign of anybody from the Indian Agency.

'Well, that's everything,' said the major. 'We'll have to hurry back now. Don't worry, we'll bring you the rest of your stuff in two days' time.'

We shook hands and thanked him profusely. They taxied the plane back into the middle of the squelching runway and a few moments later they were gone.

We realized, as the noise of the engines began to grow fainter, that we were stranded in the literal sense of the word. It was not a nice feeling, and all the more shocking for being so sudden. Pia confessed to me months later that she had never really, deep inside her, *felt* we would get to the Shavante. It was something of which she was aware in the same way as a man is aware when he crosses the street that he may be run over. Only when we stood on the airstrip with our litter of luggage about us and Biorn lying there on a rolled-up hammock, only when she looked at the knot of naked men and heard their unintelligible speech did she realize that we were here. These were to be our companions for a long time to come.

'*Are kto*,' I said hesitantly, motioning to our bags.

They hooted with laughter and then, wonderfully, replied: '*Are . . . Are kto . . .*'

They started to pick up some of our bags, and then a horseman came galloping into the clearing. He was a fair, long-faced man, who did not look like the average backwoodsman. He swung off his mount and greeted us civilly enough, but he was looking very upset about something.

'Didn't the plane bring anything for us?' he enquired.

'No, it brought us and our baggage.'

He looked at us as if he would have traded us in then and there for a sack of manioc flour.

'They are coming back,' I went on hastily. 'Perhaps you can ask them to make a flight for you then, if you need anything.'

'It's no use,' he replied fatalistically. 'They won't do it. These people are always too busy. I was hoping it was our mail plane. It has missed once or twice.'

'Doesn't it come every week?'

'It's supposed to. Every Thursday I keep a horse saddled so that I can get down here when it arrives. Sometimes it's a day or two late. Sometimes it misses a week altogether. This time it's missed two weeks. You haven't any news of it, I suppose?'

We confessed that we hadn't, and José, for that was his name, shook his head sadly.

'I didn't think you would have.'

He thought for a moment.

'Are you planning to stay here?' he asked at length.

Briefly I outlined our business and showed him our authorization to work among the Shavante.

'Well, that's nothing to do with me. The agent will see to that. We'd better be getting along to the post.'

Pia lifted Biorn into the carrying sling which she had been given by the Sherente, and there was a murmur of excited comment from the Shavante who were watching our every move. José walked his horse and led the way for us.

It was a long way. We were following a sandy track that at this season was waterlogged and heavy to walk over. From time to time we came to stretches where the water had completely taken over so that we had to wade. At one place even wading was not sufficient and we were ferried across piecemeal in the ricketiest canoe I ever sat in. The heat was merciless, and Pia was beginning to look like a lobster as she struggled along with the heavy baby. He was unusually big for his age and about a year old now. I was carrying guns, camera, tape recorder and various other items of equipment. Biorn was getting hot and hungry and beginning to whine. I felt that we were really beginning the hard way.

The ground sloped upwards slightly so that it was little by little that the post came into view. There were some Shavante hanging about, but mostly there were backwoodsmen, eyeing us imperturbably from their huts or sending their children after us to find out who we were and what we were doing. As we approached the little house under the romantic coconut palms we noticed that there were plants potted there in empty tins and that the shutters had had a coat of green paint at some time or other. There was a small half door into the porch, and over it a woman was leaning who by her appearance was clearly not like the rest of the women at the post. As we approached she was joined by her husband a small but immensely corpulent man with plump, dark features and a dazzling smile.

It was at once obvious that Ismael and Dona Sarah were in charge. José went up and reported to them about the aeroplane while we waited around in the sun. They were clearly distressed and preoccupied with the thought that there had been no plane for them. We were beginning to feel guilty about having taken up space on the plane ourselves. But they put a good face on things and showed us to our quarters in the long bleak house

which served as the post's all-purpose building. It consisted of a
verandah and three rooms or rather partitions, since their walls
did not meet the thatch roof which covered the whole. Ismael had
his office at one end and the room at the other end was a store room
which was always kept locked. The one in the middle was given to
us. A separate, one-roomed construction stood at a distance of a
yard or two from the main house and this was to serve as our kitchen.

Pia was abandoned here to feed the baby as best she could.
Lizards rustled in the thatch and scampered about the walls.
Rows of bats hung from the beams, staring blindly into the half
dark. A few rusty nails for hanging things on were driven into
the walls, but they were festooned with spiders webs, and from
them dangled the skeletons of huge spiders. She looked at the
bare mud walls and the cold mud oven with its iron warming
plate scarlet with rust, and I think her spirits sank as low as they
did at any time during the course of our field work. The baby
was tired and crying lustily but we did not have so much as a
drop of drinking water in the place.

I went over to Dona Sarah and borrowed a few essentials such
as an earthenware jar of water, a bucket and other odds and ends.
When I returned Pia was in better spirits. She flung open the
boarded window and discovered that we had the most spectacular
view of a little creek which ran into the Rio das Mortes and
beyond to the curve of the big river itself. She got some food
out for Biorn and was already deeply engrossed in creating some
sort of order in our unpromising quarters. She had also discovered
that there was a full-length door in our kitchen-cum-living
room which gave us a view right down through the post to the
cattle gate with a single, table-topped hill in the background. I
was admiring this vista with her when we saw a crowd of
Shavante come racing into the post.

For a moment at that distance it was impossible to see if they
were men or women. All we could make out were their bronzed
bodies and long flying hair. Then I noticed by the way they ran
that they were indeed men and that they were exerting themselves. Almost simultaneously I glimpsed the blue object that one
of them, in the middle of the ruck, was carrying on his shoulder.
It was our tin trunk that had been left at the airstrip! A moment
later the green trunk came into view. They were bringing them
in exactly as if they were racing logs. We knew, since they had
been carefully weighed, that those trunks weighed over 200 lb
apiece so the feat was impressive—but no more impressive than

the sight of the Indians streaming into the post in such athletic style. I understood then how the legends about the Shavante physique had got started. All over the interior we had heard stories about them. They were eight feet tall. They carried whole trees on their shoulders and could run for days on end across the waterless savannah. Their footprints were so large that they were superhuman. Their tracks pointed in the opposite direction to the one in which they were travelling. We had heard most of them at one time or other.

The facts are, as we were then able to appreciate, that the Shavante are taller than average as far as Amerindians go, but not as tall, for example, as northern Europeans. The men are magnificently built and have great powers of endurance as I was soon to find out to my cost. They brought our trunks up to the door of Ismael's house and flung them down in front of his porch with earsplitting crashes. The ends buckled and we winced. I went over to try and persuade them to bring them into our verandah.

Ismael took me aside.

'Have you got any fish hooks and fishing line?' he asked.

'Yes, I brought them some.'

'This would be a good time to distribute some of it,' he went on eagerly. 'They are desperately in need of it and I have been promising them some at the first opportunity. I have sent to Rio for it but our plane has not come and they are getting impatient. Give them some now for carrying your baggage and then they will go away for the night and let you get your things unpacked in peace.'

He turned to the Indians and his whole manner changed. He shouted boisterously at them, laughed enormously with them and kept up the clowning while he directed them to carry our trunks over. I pried them open and took out the hooks and line while displaying as little of their contents as possible—a difficult thing to do when the Shavante were pressing so close around me that they were almost falling into the boxes. The distribution worked like magic. Most of the men went off back to the village, and those that remained went in to talk to Ismael. We were happy to have some time to move in.

We were awakened the day after our arrival by the scuffing of feet in the sand outside our hut, and we could see people peering into the room around the edges of the ill-fitting window board and through the holes in the wood. From that moment on the Shavante stayed with us. The early arrivals were men on their way to do some fishing. They stopped to see what we were doing and if there was anything to be had from us. Later in the day the women came. They clustered stark naked around the doorway of the kitchen to watch Pia and Biorn together.

I got them to bring us some wood for the stove, and we cooked our first meal by the river. The next problem was to find out where to fetch water and where to bathe. Down where the creek ran into the Rio das Mortes was a tiny jetty which we saw was used for washing clothes by the women of the post. One could also swim there or further up the creek if one wished. We tried it, gingerly. The water was pleasantly clear even though it was the end of the rainy season. When we ducked under, shoals of tiny fish converged on us and mumbled around us, nibbling and tugging at any excrescences on our skin. I asked about *piranha* fish and was casually assured that there were plenty of them about. At first I did not believe it but not so very long afterwards I saw a Shavante catch a whole string of them from a position not twenty feet away from where Biorn was splashing in the water. By this time we knew that they did not come into the shallows and were apparently harmless unless excited by blood. I was dubious though on the day when I went down to have my morning dip and an alligator slithered into the water exactly where I thought I would. He backed off a little until he was round the point and in the big river itself. I raced back and fetched the heavier of our two guns, a ·222 with a rifled barrel and charged cartridge. My alligator was still there when I got back. He disappeared as soon as I fired, but I must have missed his eye for I did not kill him or apparently even wound him. The Shavante were disgusted at my ineptitude, but I was more concerned with the possibility of my alligator waiting in the shallows one day if Biorn should go down to the river.

We were still trying to find our way around the post and get acclimatized when Ismael came over to tell us that the chief of the Shavante wished to see us. It was just after midday and the whole sky seethed with yellow heat.

'Now?' I asked unenthusiastically.

'He has sent one of his sons,' Ismael explained, motioning to a well-muscled but boyish-looking youth who was standing there. 'He will show you to the village.'

'We might as well go,' Pia broke in. 'The sooner we make the old man's acquaintance, the better.'

We got some presents together and set off in single file, the young man leading the way followed by Pia with Biorn nestling against her in his carrying sling. I brought up the rear. We came down to the tiny stream which we had crossed the previous day on our way from the airstrip and once more we made the crossing in a rickety canoe. We landed further down on the opposite bank and took a path through the tall grasses.

Our arrival was unexpectedly sudden. One moment we were marching briskly through the thickets and the next we had emerged at one edge of the village. Surupredu, our guide, did not pause for a moment. He bent low and disappeared inside the first hut. We followed hard on his heels through the opening, passed between two palm screens which made a sort of tunnel at the entrance and found ourselves in complete darkness. The glare from outside barely reached the interior of the hut through irregular slits in the thatch. It took us a few moments to adjust to the absence of light, during which time we stood uncertainly right where we had entered. Surupredu had disappeared. We could see now that there was a fire in the middle of the great hut with some women kneeling by it, looking at us wide-eyed. To the right of the door opening we could make out seven or eight shiny bodies ranged on sleeping mats. There were forked poles in the middle of the floor hung with baskets and mats of all shapes and sizes.

'*Iowi*,' someone was saying to us, and I recognized a similarity with the Sherente phrase for 'Come here!' We groped our way over to where a wiry old man was patting a piece of sleeping mat beside him. Both of us lay down between the chief and his wife who got up to make room for us. The row of young men continued nonchalantly to chew nuts or repair arrows. Some of them just lay staring into space.

The chief took my hands and pressed them between his. 'Friend,' he said, and looked at me embarrassingly closely. Then he took Pia's hands. 'Friend,' he said again. It seemed as if the conversation might get bogged down at this point, so we produced our gifts. There was a fine bush knife, a box of combs, a box of mirrors and three pounds of boiled sweets. The old man took them and put them to one side, neither opening them nor referring to them. Then he took my hands again and told me I was his friend.

One of the young men shoved a hardwood club over to us. We did not know what to do with it so the chief arranged it for us

167

at the head of the sleeping mat so that we would have something
to lean our heads on. Then he motioned us to lie back as he did.
For some time he said nothing and I wondered if he had fallen
asleep. Then he sat up as if he had been stung and darted a
comment over towards the murmured conversation being carried
on at the end of the row of sleeping mats. Then he lay back
again and talked in an even, monotonous voice as if he were
addressing the thatch above his head. But this time we were
aware that everybody was listening to him.

I watched his aquiline features and the greying shoulder-
length hair and could not help feeling that the man would not
have looked out of place as a doge of Venice. Mentally I told
myself to stop romanticizing. I was simply reacting to his reputation, I thought. For this was Apewen, perhaps the best-known
Shavante in Brazil. He was thought to have led the band that
massacred Pimentel Barbosa and his companions of the Indian
Protection Service. The graves of the expeditionaries were only
a short way away in the post which now bore Pimentel Barbosa's
name. It was Apewen too who had fought off the whites for
years and incited his men to reject the presents that were
offered to them. Brazilians would time and again cross the Rio
das Mortes and leave such things as knives and salt and metal
pans on trails which they could see were in use. And time and
again they had returned to find the salt bags burst open and their
contents scattered on the ground, the knives even flung carelessly in the undergrowth. What, I wondered, had persuaded
Apewen to change his mind only a few short years ago when the
first of his men took the gifts that were left for them, leaving
arrows in exchange? Other Shavante had made peace with the
Brazilians before his village and reaped a rich harvest of gifts in
the process. Still others had never made peace and roamed the
little known lands to the north-east, where Brazilians still went in
fear of them. What decided Apewen, and why did he come down
from the Snoring Mountains, as Shavante territory is called, to
pitch his village near this lonely post?

I was wondering about the man and about the Shavante and
their fight for existence when there was a disturbance in the far
corner of the hut. Looking over I noticed for the first time that
Biorn had crawled over there and collected an inquisitive group
of Indians around him. They had undressed him, fumbling with
the unfamiliar buttons and catches in order to slip off his nappy
and see how he was shaped. The excited laughter we heard was

Shavante: Threatening the camera

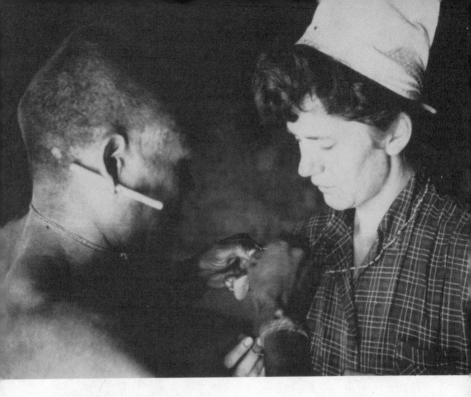

Shavante: Pia receiving a
necklace from a friend
(*above*)

Shavante bringing our trunks
like racing logs to the Indian
post (*left*)

Shavante: Boys at the Indian
post (*right*)

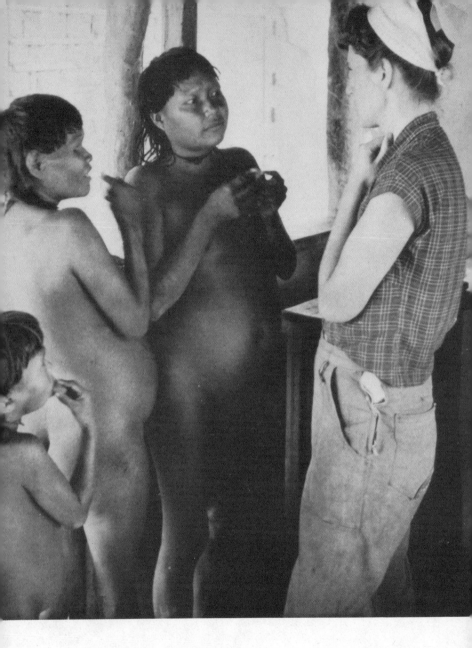

Shavante: Women with Pia
at the Indian post

an accompaniment to their discovery that he was built the same way as Shavante boys were. Biorn was clearly enjoying himself. He was crawling busily about the hut now with the Shavante making much of him and tempting him this way and that, the way one might with a kitten. Every time he headed towards the fire, Pia began to scramble to her feet but the Shavante only laughed and diverted him gently (and at the last minute) with a restraining foot.

Apewen's wife could not contain her curiosity any longer. She came quietly over and fingered our presents. The chief took the hint and opened them—not very ostentatiously but just enough to see what each package contained. He handed his wife a fistful of combs and then passed the box over to his sons. The mirrors too were handed over, even the bush knife. Only the sweets claimed his immediate interest. He sat up with one foot tucked under him and proceeded methodically to demolish about half a pound of them while we watched fascinated. All the children in the hut were around him immediately, but he crunched away unconcerned, smacking his lips and smoothing out the wrappers in a little pile. When he had satisfied his initial craving he began to distribute them around by the handful. One little girl took her half-finished sweet out of her mouth and passed it to her mother who sucked it busily before giving it to one of the men. When he had finished with it, he gave it back to the girl and she, after taking another suck or two herself, was on the point of popping the remnants into Biorn's mouth when Pia went over and grabbed him.

'I think it's time we went,' she said.

I thought so too but I had no idea of how to leave. At length I got up and in my best imitation of Shavante remarked, 'I go.' It must have been intelligible for the old man immediately replied 'Go' plus a lot else that I didn't understand. We made our way back to the post wondering how we were ever going to get to know these people.

Obviously, the first step was to move into their village. We could not hope to get any real understanding of their life if we lived at the Indian Agency and merely went visiting among them. Yet how were we to indicate to Indians whose language we could not speak and who had never had strangers in their midst before (and probably had no wish to have them now) that we wanted to come and stay in their very huts?

We had one trump card though—the baby. They were

fascinated by him, by his fair hair and blue eyes, by his clothes, his movements, everything. They came to the post well before dawn and waited for him to wake up so that they could play with him. Boys took him down to the creek to have an early morning dip with them. Women used to hide him in their carrying baskets and spirit him away from us. This had quite an effect on Pia's nerves until she realized that Shavante looked after children with the utmost care and also that they had a vivid sense of humour, even if it was of the practical joke variety.

Still, it was difficult to see how we could exploit their interest in Biorn and their curiosity about my spoken Sherente to the extent that they would take us into the village. The problem was solved in an unexpected way. The Beechcraft came back days later with the remainder of our luggage and I went to the airstrip to supervise its unloading. When it was carried back to the post we found Apewen seated imperiously on the verandah. He motioned his men to set the crates down in front of him, which they did. Then he sat back and waited for me to open them. It was a difficult situation. I had no wish, after our previous experience, to demonstrate the full extent of our resources; nor did I want to distribute everything which the Shavante might covet in one superlative handout. Yet this was precisely what they were used to. It was not ten years since the Air Force had been flying in expeditions with the express purpose of wooing the Shavante with gifts. Not five since top Army and Air Force officers had thought it worth while to fly into Central Brazil to visit the newly 'pacified' Indians in the Snoring Mountains, to make them elaborate presents on behalf of the government and to have their photographs taken in affectionate embraces with dignitaries such as Apewen. But these visits had lasted hours or days at most. Our gifts had to justify our presence for up to a whole year. It was going to be an entirely new experience for the Shavante and I was not at all sure that it was one they would welcome. I remembered the accounts I had read of how the Shavante behaved when they had first entered the huts of the expeditionaries a few years back. They had fingered everything and taken what they fancied without anybody remonstrating with them.

Acting on an impulse, I bargained.

'Here much for you,' I said, pointing to the crates. 'All'—sweeping gesture—'for you'—sweeping gesture. 'Wait . . .'

Their faces began to cloud over and we had our first experience of Shavante sulks.

'All for you,' I repeated hastily again. Then I added, 'We want a house . . . over there in the village.'

They looked incredulous so I repeated it. They began to giggle and the first crisis was passed.

'We want a house.' I was warming to my theme now. 'You build a house.' (They did not understand. Apparently the Sherente word for 'build' was not the same as the Shavante word for it.) 'You make a house. I, my woman, my child, we will come down to the house. We will make our home with you. In our house we will give you all this.' Another even more sweeping gesture. I was beginning to feel like Moses admonishing the Israelites.

They understood. They were not angry so much as mystified and not a little amused.

'You want to come and live with *us*? In the village?'

I assured them we did. But, they said, they were soon going to travel and leave the village for a long time. Me, too, I insisted. Now there were roars of laughter.

'And the woman? And the child?'

'Yes. They too.'

Finally they wanted to compromise. They would take the presents now and build us the house when they came back from their trek. No, we insisted that we wanted the house immediately. They could not understand it. Nobody in his right mind built a house just before he went on trek, and we had a perfectly good place to live in at the post, so what was the hurry? To humour us, they agreed. They would take the presents away and built us a house right now. But I knew that this was the most important concession we had to win. If they took the presents now we might or might not get our house, but it would probably not be for months. I told them, as far as my Shavante would allow, that it would warm my belly (the seat of the tender emotions among these people) to give them presents in my own house. Giving presents here was bad. Giving presents there in the village was good. After some parley they accepted it and went off. Only Apewen stayed behind, and we gave him a personal present, after which he went over to Ismael and Sarah who were accustomed to receiving him and took care that he never went away empty-handed.

I was proud of my handling of what might have been an

explosive situation, but Ismael was sceptical when I told him about it.

'They say they will build you a house!' he expostulated. 'But Indians have no sense of time. You never know when they will get around to it, and in the meantime they will plague you until they have milked you of everything you have to give them. And if you don't give, then they get angry.'

I got the impression that this was a thing to be avoided.

'Now is the time for them to go on trek,' he continued. 'They always do when the rains get lighter, and they stay away for months over there in the Roncador. They are not likely to build a house for anybody just now.'

He was right in so far as the young men returned singly all through the following days and pestered us for presents. They wanted hooks and line above all else, but they also demanded knives and matches and food and even clothing. Some of them wanted ·22 ammunition. Apewen's gifts in the period of his popularity had included two guns of this calibre and the village had acquired a third from the post.

We temporized as best we could. Pia became expert at getting the more friendly Shavante to do things for her in exchange for gifts. One might bring some wood for the stove, another fish for our supper. But many of them were too proud to serve in any obvious way. They asked for presents because it was their right, and if they did not get them they glowered or stalked off in a huff. Even if they did get them they were liable to stay around any-way, just to see what other people were getting, and from time to time they would call our attention to their bellies, grunting '*Mramti*' (hungry). We felt persecuted and bullied. Yet it was not so exasperating as the perpetual demands of the Sherente had been. The Shavante were highwaymen where the Sherente had been reduced to the status of beggars, and it is a sad reflection on human nature that most people will prefer the former. I re-membered what Wakuke had said:

'Perhaps if we too had fought, then the *civilisados* would respect us now and we would not be in our present state.'

One evening Surupredu came and asked me for my gun to go hunting. I told him that I would lend it to him only on condition that he took me with him. He was somewhat reluctant, but I insisted till finally he agreed. I thought it would be a good opportunity to get away from the confines of the post, and any-way I wanted to see how Shavante hunted. It did occur to me that it might be a strenuous business, but I reflected that I had managed among the Sherente and I was probably in no worse condition physically now.

Surupredu tapped on the shutter while it was still dark. I took the guns and slipped into my tennis shoes. I was not going to make the mistake of trying to accompany him with boots on. It was beginning to lighten as we left the post but the sun did not rise until we were already some way away and the trail had petered out completely. At first the going was through slime, which later on turned to slush and finally developed into a couple of feet of water. I was glad when the sun rose, for the morning dews and all the wetness made it a cold departure. I had not had anything to eat before I left so Surupredu gave me some of the babassu nuts he was chewing. After a few mouthfuls I had no saliva left and my throat was beginning to get dry. I stopped eating them.

As the sun rose higher I was foolishly glad that we were plodding through a watery land, for I had noted the quick springy shuffle of Surupredu's gait as he preceded me through the mud at the beginning of our excursion and remembered thinking that he would be able to keep that up all day (with the inescapable corollary—I could not). About 11.00 a.m., to judge from the sun, we came to the first stream I had seen in this desolate country and drank deeply. It was the first time we had paused since starting out. About midday I began to feel tired. It was searingly hot now and the water we splashed through felt lukewarm through my plimsolls. The scenery was as dull as I had anticipated. Hillocks of coarse grass thrust up through the endless dreary sea, and the only variation was when the stunted trees were close enough together to seem a thicket or series of thickets rather than isolated figures repeated over and over again. I kept getting the nightmarish feeling that we had already passed through this bit of country before, so featureless did I find the landscape. But Surupredu knew every tuft of grass in it.

'There,' he pointed to some grasses, to my mind utterly indistinguishable from the rest, 'a deer was killed.'

'When?'

'Last rains.' He explained to me, filling in with pantomime where my Shavante was inadequate to follow him, where the hunter had come up from, how many arrows had been fired and exactly where they had pierced the animal.

The country was no longer under water now, just oozing and

spongy. I noticed as soon as we could walk normally that the un-accustomed motion of walking through water with a 'knees-up' at every step had tired me more than I had expected. My legs started to ache. I wished my heavy ·222 was not so heavy or alternatively that Surupredu was carrying it and that I had the ·22. But the Shavante were more interested in the light gun which they knew how to handle than in the problematical heavy one of which they could only be certain of one thing—that it was heavy to carry. Surupredu was tracking eagerly now, though I could not understand how he could see any tracks at all in the watery ground let alone follow them. We darted about, occasion-ally running after invisible animals as he got more and more excited.

In the early afternoon we shot a deer—or rather he did. As we got close to it he indicated that I should remain behind as deer had very sharp ears and this one required careful stalking. Obediently I did as I was told and within a few minutes we had our first kill. I was doubly delighted. Now there would be fresh meat and we could go home. But Surupredu's appetite had only been whetted by this success. He wanted to hunt some more so he gave me the deer to carry. It was about this time that I began to get fed up with the whole excursion. The sun stayed high in the sky and refused to come down. Surupredu trotted ahead, darting off occasionally to explore a set of tracks, and I trudged behind, thirsty and tired and wishing that we were already back at the post. Worst of all we never seemed to stop. Surupredu just went on and on like a machine, and I calculated that we must have been doing this for about nine hours now.

In mid afternoon we came up with a sariema, which is of the ostrich species, only smaller. This time I was allowed to stalk it too. In fact I fired at it, determined at least to shoot once for my day's discomfort. I only succeeded in wounding it, and it was Surupredu's shot which finished it off. Now we both had some-thing to carry, and even Surupredu thought it was time to be getting back. He took the ostrich and both the guns while I staggered along with the heavier deer. I was desperately thirsty and suggested that we make for some water.

'There is no water,' was all the answer I got. 'Water is far away.'

I had feared as much. Soon, however, Surupredu stopped by some tall green grasses growing near the foot of a tree. He pulled them sharply out and quickly scooped a hole in the soft ground where their roots had been. A little water seeped into it and a film of dirt began to spread like bacteria into it.

'Water,' he indicated, 'quick.' I was too slow so he bent down himself and drank from the puddle before it had got too dirty, splashing it into his mouth with his finger tips after the manner of the region. Then he found me some more grasses and made me another puddle. The water looked bad and had a horrible taste but I was grateful for it. But the act of swallowing it down made the back of my palate and throat ache so much that I found I had simply exchanged my thirst for another discomfort. It began to dawn on me, and the feeling grew stronger as we waded back the way we had come, that I was nearing exhaustion, and this frightened me for I had never been exhausted in the technical sense before. I thought up stratagems to get some rest. My shoes were chafing and needed to be readjusted. My load was uncomfortable, and so on. I think Surupredu must have been getting tired too for he decided to lighten our loads on his own initiative. He trimmed off some of the superfluous parts of the sariema and then emptied out the contents of the deer's stomach. I was lying on a tussock, just above the water level, trying to look as if I had selected this position as the most comfortable one and not that I had just flopped down and was very doubtful indeed as to whether I could ever get up again. I could hear the bubbling and gurgling as Surupredu emptied out the deer a foot or two from my face—he too wanted to sit on a dry piece of grass—and the stench would normally have made me vomit. But I was beyond caring. I would have put up with almost anything in order not to have to move.

'*Are kto,*' said Surupredu. He was anxious to get back and get fed. As soon as I got to my feet I knew that I was literally exhausted. My pulse raced before I had taken a single step and my breath hurt my throat. A little while later I pleaded that the skin had been worn off my foot and I would have to rest it. Surupredu waited for a few moments with bad grace and then insisted that we be off again. I was desperately anxious not to lose face on this my first hunting trip with the Shavante. It was only the thought of the stories they would tell about me and my lack of fortitude that kept me going at all. I do not remember how we got back to the post or what I carried or whether I carried anything. I rather think that Surupredu ended up with the deer and I with the streamlined sariema which was less weighty. The next thing I remember is sitting in our kitchen at the post which had never looked like home to me before, and drinking till I thought I would burst.

We kept a haunch of venison; Surupredu took the rest of it and the bird back to the village. But the final irony was that my mouth and throat were so raw from my experience that for three days I could eat nothing but slops, so I never tasted the deer I had worked so hard to bring back.

'Here he comes again,' said Pia grimly.

We were having morning coffee in our kitchen with a couple of early Shavante and she was looking out through the back opening towards the clean horizon and our one hill. I peered out and saw the chief's eldest son walking unhurriedly in our direction.

'Perhaps he'll go to Sarah,' I said hopefully. She cultivated him assiduously and he had quickly discovered that he could not expect to be similarly favoured by us.

'No . . . he's coming this way,' Pia went on. We were something less than delighted. Waarodi was neither so tall nor so handsome as his other brothers, but like many an eldest son he was a person of some eminence in his own small world. He wore the mantle of his prestige less naturally than did his father. In fact, the very traits which one could not help admiring in the old man seemed to be caricatured, and therefore unfortunate, in Waarodi. He was arrogant where his father was dignified, sly where Apewen was astute. It may be that we did him an injustice and that all he really lacked was the extenuating circumstance of age. But it was a fact that Apewen could do things which were inexplicable unless one assumed goodwill on his part, whereas the ulterior motive showed in Waarodi's actions like the stones in poor soil.

'Where is the ammunition?' he demanded, strolling into the room and sitting down as if he owned the place. Today he was resplendent in a pair of white shorts which Sarah had probably given him, but he had left all the fly buttons undone so that his penis sheath could protrude.

'*Hadu*,' I replied irritably, meaning 'wait'.

'Why wait? The house is ready.'

We could not believe it.

'Our house? For us?'

He assured us it was so. We had packed a box chock full of presents for just this eventuality and now we set about gathering all our bits and pieces for the transfer to the village. I filled a rucksack with as much as it could possibly contain, took a lamp in one hand and the Primus stove in the other and set out. Pia accompanied me, carrying Biorn, and some young Shavante brought the presents and the consignment of knives which would not fit in the box. Waarodi walked pompously beside us without carrying anything at all.

Just where the path entered the village, a few yards from Apewen's hut, they had erected another one. Apewen's house was at the end of one arm of the horseshoe, ours was the second from the end. It was also much smaller than any of the others in the village, but then we did not need a large hut—or so we thought. Its thatch gleamed bright yellow beside the weather-beaten tone of the other houses. We ducked inside and surveyed our new domain.

The housewarming followed immediately. Apewen arrived with all his sons and kinsmen, who immediately fell upon the presents we had brought and started an impromptu share out. They took all the ammunition, the best knives, and innumerable packets of high-quality Norwegian fishhooks. Then they took the rest outside and shared them among the men. Women too came crowding over to see the fun, and Waarodi took charge of such articles as were to be issued to them—cloth, basins, plastic cups. For the best part of an hour there was bedlam and then everybody went away to hide his spoils and we were left alone with Apewen's sons. It seemed that we were perpetually destined to be alone with Apewen's sons. There were so many of them that they could keep a round-the-clock watch on us, and sometimes we felt that that was just what they did. With some of them we did not much mind. Surupredu, who already addressed me as 'elder brother', was a frequent visitor as was a short, plumpish young man with the difficult name of Ts'rinyoron, but we regarded them as friends whose presence placed no constraint upon us. Later we found out that they were not actually the chief's sons at all, although they slept in his hut and were raised with his sons. Most of those who were really his sons were married and no longer lived with their father. They too were frequent visitors in our hut, but they always behaved as if they were inspectors rather than guests, even if they stayed under our roof all day.

It was a busy afternoon for me when we moved into that village. I could not run around fast enough to keep the baby's wants supplied. He needed food, so I set up the Primus while Pia got out some of our dehydrated rations. Then there was nowhere for him to sleep so I had to hasten back to the post for our hammocks and blankets. As soon as I arrived back with those I discovered that he was thirsty and we had not yet fetched any water in. Before we realized it, the sun was sinking and a group of young men had assembled in one of the cleared spaces in the centre of the village.

It was no ordinary group. They were painted scarlet on their backs and bellies. Their fringes were plastered down on to their foreheads with some sort of oil and one or two of them had put scarlet paint on the coronal tonsure which every male Shavante wears. They wore hawk's feathers at their necks and clean wrist and ankle cords. Each of them had planted a speckled club in the ground behind him, so that they looked like warriors resting on their lances, and it transpired that that was exactly what they were. It was the young men's age-set gathered in all its finery for their evening council. They were men between the ages of seventeen and twenty-two who had completed the initiation ceremonies necessary for full manhood but who were not yet mature enough to be admitted to the councils of their elders. They made an impressive sight and they were conscious of the fact.

The older men did not usually assemble until after they had eaten their evening meal with their families. It might even be dark before one of their number went out to the mature men's meeting place and started calling for the others to join him. Then they would come drifting over to discuss the business of the day. I watched them go on that first night and realized that the two cleared spaces I had seen from the air were not the meeting grounds of two tribal divisions as I had supposed. The knowledge disturbed me somewhat. I had badly wanted to study a society organized in this way, and all the indications were that the Shavante would be. I got into my hammock with the uncomfortable feeling that my research in Central Brazil to date might have been following a blind alley.

It was hot and unbearably stuffy under our mosquito nets. The alternative to slow asphyxiation was to abandon ourselves to the mercy of the mosquitoes which plague the Rio das Mortes all through the rainy season. They were quite the most savage we had ever encountered. They could pierce a shirt or a sock, even a plimsoll, to drive their thirsty snouts into a person's flesh. They clustered on the mosquito netting waiting for an unwary movement which brought a hand or a leg up against it so that they could leap on their prey. Nor was it safe to shrink back into one's hammock for they were capable of biting through the hammock and a light layer of clothing as well. We despaired of finding adequate protection for Biorn. He was still in the stage of having to be bathed and changed and it was impossible to bare that pink body of his for a moment without at the same time offering it as a living target for the mosquitoes. Nor could we smear his wrists and hands with repellent during the day for he invariably rubbed his fists in his eyes and the smarting must have been more painful than a myriad mosquito bites. Eventually we discovered that he did not seem to mind the bites. He never whined or scratched them. They just mottled his skin like measles, and he scooted about on all fours happily oblivious of them. So we gave up baking him in protective clothing and larding him with pungent repellents. We concentrated instead on making sure he took his anti-malarial tablets and praying that they would be as effective as they were reputed to be.

Apewen came in to join us long before we had got to sleep. The old man apparently intended to sleep in our hut that night, and I was pleased, for the implications of his acceptance and patronage of us boded well for the future. But his presence was a mixed blessing. He was a light sleeper and spent the early part of the night rustling about the hut, sometimes going outside for a snack (which involved a cataclysmic swishing of thatch as he went out and in), sometimes calling to others outside and carrying on conversations with them. When he finally stopped grunting and chewing he fell into a sound sleep, lulled by his own tempestuous snoring.

I felt as if I had only slept for a few seconds when I was awake again. Apewen was gone. Perhaps it was the sound of his going that had woken me. I could hear a noise like the scuffing of innumerable bare feet in the sand outside our hut. Pia was wide awake by this time too.

'What's that?' she said hoarsely.

'Sounds like a party of them going somewhere,' I replied, trying to sound authoritative, as if I had good grounds for believing that parties of Shavante normally wandered around in the middle of the night. By now we could hear low voices and people clearing their throats. It sounded as if they were gathering right by our entrance and it was clear that they were all men.

I swung out of my hammock and dived for my flashlight. As my fingers closed over it there was a strange thumping from about five yards away. I crouched there in the dark holding my torch and listening harder than I have ever listened before. Then the

179

singing started. I went out into the dull night and guessed rather than saw that Apewen was sitting outside our entrance. A circle of youths with linked hands was doing a rhythmic stamping dance in front of his hut and accompanying it with a guttural barking chant.

'*Tirowa*,' Apewen said to me, adding by way of explanation, '*He'wa*.' I learned later that *Tirowa* was the name of an age-set and that it was this age-set who were the people currently in the *He* or bachelors' hut. When they had finished their song they formed up in a single file and moved on to the next hut. Gradually they moved round the village, dancing and singing till their voices grew hoarse and they began to miss a hut here and there. We slept again with their strange chorus in our ears.

Next morning Surupredu came to beg for ammunition.

'But I have just given you some,' I protested, 'and there are boxes of it over in Apewen's hut.'

'We are going far. We are going on trek.'

'Already?'

'Yes,' he counted on his fingers. 'One sleep. Two sleeps. Then **we go** on trek.'

This was a real blow. No sooner had we managed to get into a Shavante village when it moved away from us. I was not sure whether we would be welcome if we accompanied them, and furthermore I was not sure whether it would be wise for all of us to do so. If it had been rash of us to bring Biorn to this outpost in the wilderness, then it would be doubly so to set out into the Snoring Mountains with him. After a day of discussion we decided that I would go with the Shavante if they would have me and that Pia and Biorn would remain behind. I was not altogether happy about deserting them there but there seemed to be no other reasonable solution to our dilemma and I took comfort in the Webley which was in her possession. Pia, on the other hand, was unhappy that she was not going to accompany the Indians in their wanderings. After all, she reasoned, Shavante babies accompanied their families so the circumstances of trekking could not be so inclement. But then Shavante babies had Shavante parents to look after them for one thing, and for another they occasionally died.

The next problem was to find myself a portable food reserve to take out on trek. I had neither the time nor the skill to forage for my own food as a Shavante would. In any case, I needed something to offer my hosts in return for their hospitality. The obvious solution was a sack of manioc flour. Shavante loved hard, crunchy foods and they especially loved manioc flour or *upadzu* as they called it. I knew this would be the most welcome thing I could bring them. The difficulty was to get hold of some. Every Thursday José saddled his horse and the tiny, forgotten community went about its humdrum tasks with one ear cocked for the sound of 'their' aeroplane. Every Thursday it failed them. Food began to run low at the post. Clearly something had to be done. Eventually Ismael sent some men up river to see if they could buy supplies, and they were commissioned to bring back a consignment of manioc flour for us.

The Shavante, however, would not wait. It hardly rained any more and little by little the floodwaters were beginning to dry out. They felt it was time for their spring migration. The good things that I promised them from that plane when it came were no longer so attractive as their own ranging grounds beyond the river, and anyway they could manage very well without any of my gifts. The best I could do was to get an assurance from Apewen that he would direct somebody to take me out to the villagers as soon as my supplies arrived. They wanted to have my manioc flour even if they did not want me.

So it was only a day or two after we have moved into the village with such high hopes that we started to pack our things and carry them back to the post again. I saw no advantage in Pia's staying down there alone with Biorn. The next morning the whole Shavante band moved off. I did not see them go, for they wandered off in different directions and I was not at all sure who was going where or with whom. I even discovered that some of them were not going at all. Apewen and his wife and infant son stayed behind together with one or two others who were too sick or infirm to travel.

An unnatural calm settled over the post. We drank endless cups of coffee and listened to the dreams. In the absence of news from the outside world dreams became of paramount importance. A vivid dream was the only novelty of the day. It served as a talking point, news bulletin and indication of the future. Sometimes competing dreams were told and analysed and measured against each other like fighting cocks.

I had rigged up my portable radio transmitter, and when the vultures did not sit on the aerial or the Indians take pieces of it for fishing line I could even operate it occasionally. I had asked the radio operators in Shavantina and Aragarças to tune in to my frequency now and again at 7.00 p.m. but I never managed to make contact with them. I decided that perhaps I was not sufficiently experienced in its modulation or alternatively that our watches were all wrong. It could receive, but the news bulletins about international affairs were meaningless to the inhabitants of the post. They wanted news of the aeroplane, of Shavantina and of the likelihood of someone from the Indian Protection Service flying out to them with their pay. These were the things they dreamed about and very soon we too found the dreams and their interpretation more real than the newscasters. I did not even tune in any more.

At last the canoe came back with provisions and manioc flour for me. Apewen arrived to share what was going almost before we ourselves heard the splash of the paddles.

'Now I will take much *upadzu* to your people,' I told him.

He did not look overjoyed at the prospect but cheered up a little when we measured him out a bag full too.

'Tomorrow,' he announced suddenly, 'Pahiriwa will take you,' and he pointed towards the columns of smoke discernible on the horizon. 'He will take you to our people.'

9

In the world of the nomads

From the start I had my misgivings about that journey with Pahiriwa. I did not know him very well but I did know that he did not really want to go on the trip at all. I did not fancy the idea of being stranded in the middle of nowhere with a Shavante to whom I could barely talk. Still, I reassured myself, I had distinctly heard his father say that he would take me out to the hunting camps, and I was still new enough among the Shavante to imagine that a chief's word, if not law, was something in the nature of a covenant which could not lightly be put aside. The test, Pia thought, was whether Pahiriwa showed up at the post on the morning scheduled for our departure. If he doesn't want to go, she reasoned, then he will simply not appear. But there she was wrong.

He appeared all right and accepted the coffee and cakes that we offered him in the same regal manner that he accepted everything else. I could not help admiring Pahiriwa's manner. He never made up to us and never begged. He did not even unbend among his fellow Shavante. He rarely laughed and when he smiled his handsome but heavy features looked quite medieval in their condescension. He never looked round in the quick, alert manner that characterized the other Shavante but always turned his head slowly as if he had a stiff neck. He rather intimidated me.

When we carried my gear down to the makeshift canoe, I felt as if I were the dragoman and he the travelling potentate. Nor did the inhabitants of the post who gathered silently to watch us go do anything to dispel this impression. I was feeling distinctly uncomfortable as we floated round the bend and I lost sight of Pia, still waving faithfully from the best vantage point she could find.

I had prevailed upon Ismael to let me have a couple of horses, one for the manioc flour and one for myself. I had not really expected to be granted the second animal but pressed for it anyway. After my experience with Surupredu out hunting I wanted to make sure that I could keep up with the Shavante. Two of the men from the post had ridden them downstream to a point where the river was narrow enough for the mangy animals to be able to swim across, and we were now on our way to meet them and exchange means of locomotion. I had at first objected to this arrangement as over complicated. Why could not Pahiriwa and I take the horses down to the crossing place and go on from there ourselves? City people are prone to imagine that they can work things out more logically than the locals. Patiently they explained to me that the horses were in such a shape that even at a comparatively narrow part of the river they would never make it across to the other side without a canoe to help them over. That meant a canoe had to go too and while there was a boat involved it was best that the baggage should be carried in it and not on the animals.

River travel by dugout canoe is, in my opinion, the most uncomfortable form of locomotion that it is possible to imagine. When travelling in the sole company of Shavante, who are appalling watermen, it can also be mildly hazardous. These Shavante had, after all, neither built nor used boats until they were permitted to borrow the ones belonging to the Indian Protection Service. They thought of rivers purely as obstacles and traditionally they swam across them, holding on to a tree trunk, or built rafts for their families to float across. I enjoyed my excursion that day because I knew that we had only two or three leagues to go—however far that was. Although our departure had been delayed, as departures always are in the backwoods where packing is so complicated, the early morning freshness still had not worn off. The water splashed agreeably and though there was none of the wild life of the Amazon jungles to delight the traveller, the Rio das Mortes had its compensations. Occasional brightly coloured birds cawed their way across it. Fish jumped. Alligators eyed us stonily. Flamingoes, statuesque on a sand bar, disdained the futile shot that we fired at them from our rocking boat, rose slowly into the air and flapped away like flying bedjackets. Three otters came swimming towards us chattering. I was torn between the desire to take a pot-shot at them and the feeling that it would be a crime to disturb the smooth perfection of their motion through the water. Pahiriwa resolved my aesthetic problem for me by grunting 'pig' and swinging the canoe so quickly that I nearly fell into the water.

I have never met a Shavante who would not drop everything to go after wild pig and Pahiriwa was no exception. We beached

Shavante: Suwapte coming
back from the hunt with a
captured peccary

Shavante: Men and boys
start out on trek (*above*)

Shavante: Women and girls
prepare to go on trek (*right*)

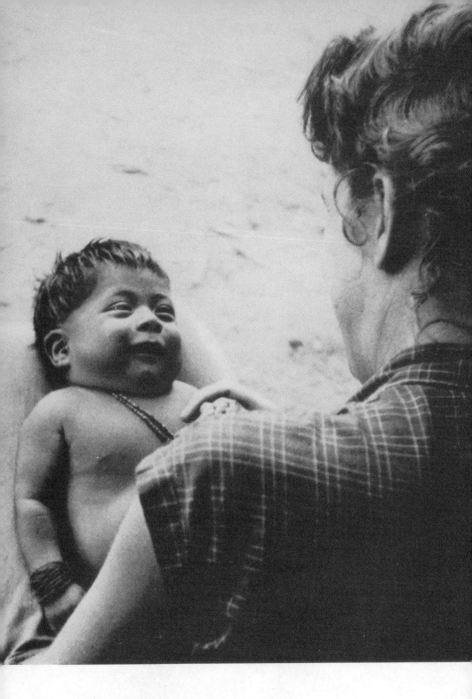

Pia holding a Shavante baby

the canoe, sloshed barefoot up an oozy bank and scampered through thickets and around creepers; Pahiriwa tantalized by the snorting noises which seemed to be all around us, I more intent on keeping the thorns out of my bare feet. We never did see those pig but Pahiriwa redeemed the expedition by sighting lengthily with the ·22 and bringing down a sitting mutum. I have often felt that those people who are shocked by the idea of shooting a sitting bird would do well to spend a month or two in Central Brazil with a rifle and a necessity to provide their own meat. A mutum is the size of a small goose, so we were well pleased with ourselves when we edged the canoe into the spot where we could see our horses switching idly with their tails over their sleeping riders.

By the time we had swum the panting horses across the river and loaded up the pack animal the sun was at its zenith. The men from the post assisted us in the operation and it gave them plenty of opportunity to exercise their dour sense of humour at my expense. They even seemed in quite a hurry to get into the canoe and get going although they had that long battle upstream to look forward to. They left me in no doubt as to the reason for their eagerness. Their little spell of discomfort was a bagatelle compared with what they were convinced I was about to go through.

'Oh, you will suffer, *senhor*,' one of them said with a broad grin. 'I have heard that the gadflies are terrible and that they do not give you a moment's peace from dawn till dusk.'

'Yes, and living with the Indians is no joke,' chimed in his companion. 'You won't be able to stand it physically.'

'It's out on these hunting trips that they kill off the people they don't like. They don't do it here in the village because they feel the Service might get to know about it and stop giving them presents.'

By now they were in the boat, pushing off.

'*Vai ser dureza mesmo*,' called one of my Job's comforters as a parting shot. The expression, meaning literally, 'It will be hardness itself' struck me later as being the best description of life among the Shavante that I had heard and certainly the pithiest. I sometimes felt, when I was alone in a Shavante encampment, like a limpet on a rock. The whole universe seemed to be made up of hardness. The ground we slept on was hard. The food we ate was hard and had to be torn apart by fierce teeth or held captive in the mouth and sawn off with a knife. The glare was hard by day and the cold was hard by night. The ways were hard to travel, hard underfoot as we approached high summer and hard to tear a path through.

At the moment though I was in no mood to appreciate epigrams about the Shavante. I was apprehensive and Pahiriwa was sulking. It was already late, he said. Why did we waste so much time? When he did this journey alone and on foot he was almost there by midday. Why was there no horse for him? I thought to mollify him by suggesting that we rested in the shade and ate some of my provisions until the sun got cooler, and to shame him by saying that I had never heard of a Shavante *man* who needed a horse to get about in his own territory. Pahiriwa showed his displeasure with both these arguments by stumping off up the trail and leaving me to follow on with the horses as best I could.

This was not nearly so easy as it might sound. In the close country that stretches away from the wild west bank of the Rio das Mortes a man on horseback is at a considerable disadvantage. As for the packhorse . . . well, I had plenty of time then and later to regret the packhorse. But how else could I have carried all that manioc flour which was not only my passport to the Shavante but my only food reserve? That was little consolation to me as I urged the reluctant animal over what I hoped was the trail Pahiriwa had taken. Sometimes it was easy enough to see where he had passed, but at others I was in an agony of doubt either because my unpractised eye could make out no spoor at all or, with increasing frequency, because such large numbers of Indians had passed that way so recently that I could not make out which were his tracks. At last we blundered into a really thick piece of jungle and the packhorse stumbled over a tiny shelter, built like a sentry box just off the trail, almost before I had seen it. I peered to the right and to the left and could see the semi-circle of shelter groups with here and there a few bones or the remains of a long-dead fire. This was one of the Shavante hunting camps which meant that we were still on the right trail. It was an eerie feeling to stand alone among those empty shells and wonder how much further ahead the Indians were. How much more so, I reflected, must it have been for the earlier expeditions that prowled nervously through this part of the country, which even today is undisputedly Shavante territory. There was no sign of Pahiriwa, so I mounted hastily and hurried on. Now we passed through one of these camps every few hours so at least I could be sure we had not lost the way.

I caught up with Pahiriwa in the late afternoon. I rode out from among the trees, one eye on the load jazzing crazily on the irritating pack animal and the other on the faint trail which I was praying was his. I found myself among the shelters of another camp, built out in the open, and there, to my intense relief, was Pahiriwa, lying flat on his belly in the middle of the path. He was trying to get a drink from a water hole which the Shavante had dug but the hole was old, and the white, misty water, looking like ammoniated quinine, was a long way down. I dismounted, took out one of the mess tins which I always carried and handed it to him. Not only did it extend his reach but it held considerably more water than the palm of his hand. For the first time that day, Pahiriwa looked pleased.

We sat down in the miserable shade provided by one of the derelict shelters and ate some of our provisions. I was carrying a bag of the only picnic food which is known in the interior: *fritada* or toasted manioc flour with little bits of meat in it, all fried up together so that it will keep indefinitely. Pia had cooked what we fondly believed to be enough for me and my guide for at least two days but I could see from the way Pahiriwa kept digging his big hand in and tossing the stuff into his mouth that we would be lucky if there was any left over for supper.

That night we roasted the mutum and ate it with its skin charred and crusted with earth. I found it delicious, but then I have always liked well-cooked food. We lay down to sleep with full bellies, and Pahiriwa seemed almost reconciled to me and my journey. Next morning he was as surly as ever, for it took me a long time to track down the horses that had been cropping the grass all night. By the afternoon of the third day he was in a foul temper.

It was, therefore, a great relief when the jungle gave abruptly on to rank grassland and I saw the shelters grouped like pimples on the savannah with their swarms of children around them. We had caught up with the band at last. But I soon wondered if I had not come out of the frying pan into the fire. Pahiriwa stalked off to his mother's shelter without a backward glance. He did not even offer to hold the load while I took it off the packhorse —things which are very difficult for one man to do on his own. The men were all out somewhere—hunting probably—the women were afraid of me, and the children just stood and gaped. Shavante custom, so meticulous in most matters, provides only for hostile relations with white people, so nobody knew how to greet me or what to do with me. I sat outside the shelters in the hot afternoon sun, feeling conspicuous and foolish and trying my best not to look it.

It was Suwapte, the chief's son-in-law, who rescued me from my vigil. He came back at dusk and took me to his shelter. He was careful to exact half of my store of manioc flour too. That night I sat for the first time in the men's circle. There was no moon and I could only distinguish the humped shapes of the 'mature men' as they sat on their deerskins to my left and right listening to the impressive barking oratory of the speakers. A background of comment and interruption flickered round the circle, intense and at the same time indeterminate like distant artillery fire. Now and again one of them lit a bunch of dry grass to warm himself. It would burst into flame like a flare, making the men's eyes glow in the darkness and giving their long hair and fine shoulders an unusually barbaric and aggressive look.

I realized that it was Pahiriwa speaking. In his deep, sarcastic voice he was giving a minute, detail by detail account of the days we had spent on the trail, of what we had said, what we had done, what we had seen (and, of course, what I had failed to see). Now there were guffaws from his audience. I was nudged and slapped on the back by my neighbours. At least, I thought, if I am accepted as the local buffoon, I have a well-defined role in this community.

In the weeks that followed I found that I identified myself more and more with this little group of Indians, wandering isolated in the wilderness. Suwapte adopted me and acted as my mentor. I kept my things in his shelter and ate and slept there. I found sleeping rather a problem at first, for the tiny shelters which Shavante build when they are on trek can only just accommodate the people they are intended for, and then only if they lie like sardines in a tin. Not all the inmates of a shelter can lie on their backs at once or the people sleeping at the end of the row are pushed into the thatch. I had perforce to lie intertwined with the other sleepers and was kept awake by the jarring of their knees and elbows. Shavante seemed to be able from long practice to fend for themselves in their sleep. It was worse as morning approached, for the dry season is also the cool season here and we were about 3,000 feet above sea level. It got colder and colder and the sleepers huddled closer and closer together. Sometimes I would find my blanket being shared simultaneously by people on both sides of me which made my middle berth somewhat less than comfortable. Strangely enough the women were more aggressive sleepers than the men. An occasional heave would establish my right to a few square inches of space if I was being crowded by one of the warriors but I was sometimes so steam-rollered by the women that I gave it up and went outside to squat by the fire and wait for morning. Most of the householders gathered there long before dawn anyway. It became too cold for comfort in the shelters, so they would come out and sit cracking nuts and blowing into the embers, stretching themselves and waiting for that marvellous dawn hour which was so good for hunting or travelling or just bathing.

Trekking was a dirty business too. Shavante often cleared the ground by directing a ragged bush fire over the site they planned to move to and then they camped in the ash so that living in their shelters was like living in a chimney. Everything in them, including their inhabitants, got covered in grime. Often the only water to be had was from water holes so that washing had to be restricted to taking a mouthful of it and squirting it in a thin stream over one's hands. For the rest the Shavante titivated themselves by chewing nuts, spitting the oily juice into the palms of their hands and anointing themselves liberally on the body and hair. I got used to the smell in the shelters and to the offal that littered their entrances where the cooking fires burned. I found that I did not much mind that the starchy roots which were the basis of our diet were pulled out of the fire, covered in earth and ash and thrown casually over to us in this state. But it took me a long time to get inured to the fact that they always plumped into that patch of ground just inside the shelter which was invariably covered with garbage and mucus, to be retrieved, dusted off and eaten. Still, though it turned my stomach, I was glad to have food tossed over to me just the same as the other men. It made me feel accepted.

I still had one besetting fear—that I should get lost. I was fully aware of my own deficiencies as a backwoodsman, and when I went out with the Shavante my attention was usually occupied with other matters than trying to memorize landmarks in a featureless country. Moreover, I knew that the Shavante could not understand such obtuseness on my part. They thought it incredible that a grown man could be so helpless, much as New Yorkers seem to have trouble in understanding that anyone who is not mentally deficient might have difficulty finding his way about New York. It was, therefore, quite possible that the

Shavante might lose me out of absentmindedness rather than malice aforethought and I always stuck to somebody or other like a leech when I was out with them.

On one particular occasion I could not do this. It was one of those hot, insect-ridden mornings when nothing seemed to be happening in the camp. I had scrambled down a rock face to get at the only pool of water in the neighbourhood, washed and returned to my shelter for breakfast. This consisted invariably of a plate of manioc flour and milk. I always made a mess tin full for myself and one for some other member of the hut. Today I had handed a tin full of cereal to Suwapte who was nursing an injured leg and was preparing my own, relishing the thought of it (food was getting to be the high point of my day too), when a shrill hunting call came from just outside the encampment.

Bedlam broke out. Men shouted and women screamed. Two or three people came bounding out of their huts with bows and arrows. I ran outside to see what was happening and was met by shouts of

'*Uhe! Uhe! Bare uhe da!*' (Pigs! Pigs! Quick, after the pigs!)

For a moment I hesitated, thinking of my uneaten breakfast already anointed with milk. Suwapte came bounding towards me, his hurt leg forgotten, his bow and arrows in one hand.

'Come on,' he shouted. 'Come after the pig!'

'Where are they?' I temporized.

'Close, very close,' he called over his shoulder, giving the words an upbraiding inflection that cast aspersions on my lack of enthusiasm. My conscience got the better of me. An anthropologist could not surely think about such mundane things as breakfast when 'his' people were streaming out on a pig hunt—especially when the people virtually lived for pig hunts.

I ran back to the shelter, collected my gun and went to join the hunters. To my surprise I found them assembling a little way from the village under a tallish tree. Surupredu was up in the tree repeating the hunting call for pig at intervals. The others were standing around and there seemed to be no sign of pig. I think I had half expected to see a flock of pig browsing their way through the undergrowth. I had certainly hoped to kill a couple as quickly as I could and then return for my breakfast. Instead I fell in glumly as the file of hunters made their way down a slope and off into the bushes at an excited shuffle. There were tracks everywhere and now again there would be volatile discussions as to where we should go next. We appeared to be tracking about six animals simultaneously. I heard tapir mentioned and deer and more and more peccary, but in the meantime we drew further and further away from camp. Nor did we come up with the animals. Sometimes we were apparently so close to them that the men scattered and went careening through the undergrowth in full cry. We tore through thick jungle where a watercourse meandered. We ran across miles of open savannah. The sun climbed impudently higher while our efforts went unrewarded. The hunters were beginning now to talk in terms of staying out all night to keep up with the animals. My shoes were sodden and pinching. I was very hungry and did not look forward to the idea of spending a night out on the steppe without my blanket. In short, I was tired of the whole excursion.

I was in this frame of mind when the party killed a deer. This created a problem. Nobody wanted to go back with it but it would obviously burden down the expedition. I saw my opportunity to bow out with honour and offered to take it back to the camp. The hunters were overjoyed.

'Who is coming with me?' I asked none too hopefully.

'You.' . . . 'You.' . . . 'You alone,' came the replies. 'Why go with you?'

'To show the way,' I admitted.

The men were nonplussed, then angry. Why show the way? I knew the way . . . everybody knew the way. It was dead straight all the way back to the camp, and besides we had left tracks. . . . They indicated with their arms, like boastful anglers, what sort of tracks we had left. I realized that to hesitate now would be to lose more face than I was prepared to. I said I would go back alone. At once they were all smiles and eager to be off.

I started back along our trail, feeling rather like William Tell. The deer's tongue was lolling all over my back and I was getting crusted in its blood. I thought of the labour it would entail to wash my clothes in the scarce water of this region and so I took them off and carried them too. For most of the way the going was indeed relatively easy and I could follow our broad trail without difficulty. In mid afternoon I plunged confidently into the jungle, but the trail I followed inexplicably petered out. I tried again with no better result. This time I turned round after my failure and retraced my steps. It was a relief to find that I could at least get back to the place where I had entered the jungle. There I took stock of the situation. I noted the position of the sun, the general direction in which I wanted to go, the tracks we had been so interested in that morning—everything which might help me to retain my bearings. Then I tried a third time. I got to the stream we had crossed and correctly worked my way through to it again as it doubled back on its course. Then I navigated as best I could in the direction I wanted. The trails gave out and I had to hack my way through the undergrowth. We had not done that when we passed that way in the morning. Wearily I made my way back to the stream and tried to use the old trick of wading down it until I found the point at which we had crossed. I sank deep into the mud at every step, weighed down as I was by the dead deer. Then the bottom sank away without warning and I disappeared under water, deer and all. With great difficulty I extricated myself and floundered out on to the bank, getting severely cut and slashed in the process for here the jungle was really hostile. The tall grasses were razor sharp, the close set branches spiny and unyielding. The mesh of trailing strands joined hands in whichever direction one wished to go, barring all egress with a malignant thoroughness that it was very easy to take personally. At this, of all times, I discovered that I had lost my hunting knife. It must have jerked loose from its sheath when I submerged in the steam. So I sat there, bleeding and tired. I was quite definitely lost.

I remembered the most important rule under such circumstances—try to keep calm. I did not go blundering frantically through the undergrowth and exhaust myself by travelling round in small circles. On the other hand, having kept calm I was not very sure what to do next. My calculations had not helped. The sun was now sinking pretty low and was correspondingly less useful as a directional guide. Anyway it was rare in this closed world of the forest that I could get a glimpse of it and impossible for me to steer by it. I had searched for tracks and found so many that they were of no use to me. The more I thought, the less I knew what to do and the less I knew what to do, the more difficult it was to go on observing the golden rule of keeping calm. I remembered the stories of people who strayed away from their boats on the Amazon and who died from exposure or fright or exhaustion a couple of hundred yards away from companions who could neither hear nor find them. This was not the Amazon jungle but I derived small comfort from that fact. It was just as capable of swallowing me up unless I acted coolly. Perhaps, I thought, I ought to sit still and wait for the Shavante to find me, but I did not know whether they would bother to search, and if they did, it might be a day or two before they realized that I had not simply gone off on my own business and perhaps another day or so before they tracked me down. No, that was an unpleasant prospect. Preferring action to inaction I tried cautiously to steer in the direction I wanted to go without paying any attention to tracks or trails. It was very slow and uncomfortable. Eventually, though, I got out of the jungle and into forest country which was just as unfamiliar but where I did not have to fight for every yard of progress. I hooted from time to time like the Shavante do, but was answered by no hail.

I do not suppose that I was lost for more than about an hour, but complete spatial disorientation plays tricks on one's time sense. I felt I had aged at least a week when I struggled on to what was obviously a well-trodden trail and saw the most beautiful children I remember—two little Shavante who had wandered away from the village to play. They were goggle-eyed at my appearance, and I realized that not only was I dishevelled, I was also naked. Not that Shavante mind particularly about nakedness but they feel that for a man to be seen without his tiny penis sheath is the height of immodesty—and I was not wearing one. Hastily I put on my slacks and accompanied the children back to camp. They made for home immediately themselves when they saw that I was bringing back meat, so I was able to re-enter the encampment without it being too obvious that I was being shown the way. I could not cover up the evidence of my battle with the jungle though. When I saw myself in a mirror, I realized that I looked as if I had been whipped all over.

The Shavante thought this was hilarious. They could hardly concentrate on skinning and cooking the deer for their interest in my misfortunes. The women came from all over the encampment and besieged our shelter demanding to see the welts. The men noticed immediately that my knife sheath was empty and demanded to know what I had done with the knife. I told them and they were even more amused. My escapade in the jungle was the main topic of conversation that night in the men's circle, and the whole story had to be retold the following night because the party of hunters I had set out with did not get back till the next day and they wanted to hear the story too. The hunters, it turned out, were cross with me.

'Where is the venison for us?' they demanded of me when they got back. They were loaded down with pork, but let that pass.

'The women took it and distributed it,' I replied.

'What for?' they expostulated angrily. 'Why did you give it to the women?'

I could hardly explain to them that I had not the slightest idea who to give the animal to, that it had been torn out of my hands by the women of the shelter I was in, and that I had presumed they were the people who would normally take care of booty from the hunt.

'You have a lot of pork,' I countered, deciding that attack was the best method of defence. 'Why do you need venison?'

They were not mollified.

'There is lots left,' I continued less hopefully. 'Your women will have some for you.'

'*Meh!*' they replied, using a Shavante expression (pronounced to rhyme with that delightfully absurd English interjection 'Heh' as in 'Heh! Heh!') which indicated a nice blend of irritation, scorn and disinclination to continue the conversation. But even the hunters cheered up when they heard about the trouble I had bringing the deer home. They laughed uncontrollably and called on me to retell the exact circumstances of my journey again and again. It became my party piece, establishing me, I gathered, as the local jester. They were still chuckling about it when we met up with another band from Apewen's village, and the first night that the two camped together I had to tell the whole story all over again.

'Our band', as I had now come to think of it, was trekking slowly into the Serra do Roncador. We were only a small fraction of Apewen's village but we were not as isolated as we appeared to be. Every day we watched for the tell-tale smoke which marked the progress of the other hunting bands. Even lone travellers or small parties 'made smoke' as they moved across the landscape so that we were able to keep track of every peaceful Indian within a wide radius. Similarly our far-ranging hunting parties always brought back full details of the tracks they had seen, both animal and human, while they were out, and these reports were eagerly discussed in the men's circle by night. Tracks to the south and east were innocent tracks, tracks which could be given names.

'That was brother-in-law going to collect fruit', or 'That was uncle when he killed pig during the rains'.

But tracks to the north and west were more ominous. Some of them we had made ourselves—but what of the others? There had been no smoke from that direction and Shavante who did not make smoke were Shavante who did not wish their presence to be noted. How old were the tracks? How many people? Whither did they lead? These were burning questions which I heard debated over and over again, and gradually I began to learn something of the relations that existed between Shavante communities.

These can be summed up very easily—they were bad. Apewen's Shavante were on bad terms with the Shavante up river who had established their base camp near the Catholic mission station of Santa Terezinha. But the Santa Terezinha Shavante were apparently frightened of Apewen's group. On the other hand, Apewen's community was also on bad terms with the 'wild', uncontacted Shavante away to the north-east at a place which the Indians called Marawasede. More than that, they had a healthy respect for the people there. That was why our community was running out of bows. They made their bow staves only out of a certain wood which grew in abundance in the territory controlled by the Marawasede people. Occasionally they would trek over there in force and collect what they wanted, but individuals and small groups would not risk it. If they broke their bows or bartered them away to passing airmen then they would borrow one from a kinsman until such time as the whole village went to get new ones.

So, although we were in the heart of Shavante territory, the Indians themselves were on their guard. Against such a background of wariness, it was perhaps natural that I should have been apprehensive the day I saw a young warrior standing quite alone on a bluff, looking just like the Scout who is so familiar an apparition in Hollywood westerns. The whole village was on the move that day, and the others had also seen him but they gave no sign of it. As we got closer to him I realized that this was because he was from our own village. All the same, here he was an outsider. I passed him, as did the others, without a word. When we made camp he appeared and went into the shelter built by one of his kinswomen. All that day he lay there, accepting food and talking of trivialities. Only his closest relatives asked him his business. The rest ignored him; but the whole village buzzed with rumours. Some thought that the wild Shavante away to the north had come out on the river and killed many white people. Others had heard something about the bad chief Eribiwen to the south.

It was not till that night that the messenger strode importantly into the men's circle, leaned on his great bow and began his speech. The Shavante upstream were angry, he said. They had killed a chief—they had killed a chief. (Shavante speeches are very repetitive.) They had killed Eribiwen when he visited their village. Eribiwen's people were very angry. War parties had been out—here he gave an account of the killed and wounded, showing where the clubs had struck and exactly where the arrows had gone in. The white men were angry, he went on. Shavante had attacked the mission and the white men had sent aeroplanes to take away the wounded.

I could only understand half of it, and many of the references to places or people that I had not yet seen were lost on me. Yet out of this welter of talk I pieced together the story of an Indian skirmish which apparently filled the Rio newspapers with wild tales of Shavante on the warpath and even brought rumours into the foreign press about missionaries being killed and, incidentally, one anthropologist and his family having disappeared!

At the time it affected Pia and the people at the post much more than it did us who were far away in the hinterland. It was they who first got the news of the fighting, not from any Shavante messenger but by air from Shavantina. José went galloping out on his horse at the sound of the aeroplane engines and the peons in the agency went wild with anticipation, only to be disappointed again. It was no Air Force plane flying in with supplies but one of the two civil machines belonging to the Central Brazilian Foundation's base in Shavantina. The pilot nervously kept his engines running until José arrived.

'Is everything all right here?' he asked.

José was obviously mystified.

'Are the Indians at peace?' he insisted, looking nervously at the two or three who had come down to the airstrip. When he discovered that nothing untoward had happened in these parts he told a garbled story of the murder of a Shavante chief which had led to raids and killings upstream. He consented to come down to the agency and explain the whole situation to Ismael, and, of course, old Apewen came hastening up as soon as he got wind of what was afoot. There was some talk that Apewen should be flown back and taken by river to act as mediator between the conflicting groups, but he categorically refused. In view of my subsequent understanding of Shavante politics, I am not surprised that he did. In fact, he was nervous about the possibility that the Shavante from upstream might come and raid his people.

He at once despatched a messenger to tell his cohorts what had happened—the man from whom I heard the news—and, furthermore, to tell his sons that he yearned for them in his belly and wanted them to hurry back to the village.

For days the possibility of a Shavante attack was in the minds of everybody at the post. Apewen came frequently up from the village and used to stand in Pia's porch looking longingly at the columns of smoke far to the other side of the river where his people were. In time he made Pia thoroughly nervous too. They carried on a laconic conversation about the situation, a conversation punctuated with Apewen's emphatic exclamations of '*Pipa-di*' ('frightened'—or 'be careful'). Then he would go stumping back to the village, leaving Pia alone with Biorn in the long, isolated hut by the river. She did not get much sleep. The squeaking and rustling of the bats kept her awake where she had ignored it previously. All the night noises of the river and the forest now craved explanation. She thought ruefully of the line of rusting guns hanging in the storeroom next to where she slept. They were the guns which Pimentel Barbosa's men took with them on their expedition—and had kept locked away. It was not in the tradition of the Protection Service that they should fire on Indians, however hostile. Perhaps those same guns would be locked away the next time the Shavante attacked—and yet she knew that if they did come, there was no defence. There was nothing to do but pray that they would not. Each day she felt relieved when the sun rose and she could take Biorn down to the river and pretend that everything was normal.

Up where I was in the Snoring Mountains, I was blissfully ignorant of this complex series of events. But as I listened to the impassioned speeches that were made on the issue raised by Apewen's messenger, I began to wonder whether I would not soon be involved in a war party.

Fortunately no immediate decision was taken. Had I known the Shavante better then I would have realized that they probably would take weeks to agree on their plan of action or inaction. But the fact remains that I worried. In the next few days there were often puffs of smoke from the south, and in the evening there would be a fresh face in the men's circle to bring us the latest rumours. There were no more reports of fighting though and there seemed to be less and less talk of retaliatory expeditions. For some reason or other my group had allowed their bellicosity to evaporate and they were all for continuing their trek unless they received an urgent summons from the old man. Even the messengers had less to say about Shavante hostilities. Clearly they brought no fresh news on this topic and there was a limit to the amount of discussion which could be extracted from the original reports of the fighting. Instead they brought news of Biorn.

I understood that he was now known as Sibupa after one of Apewen's sons. Similarly the old man had bestowed his own name on me and decided that Pia should be named after his present wife. I felt honoured and only wished that Apewen's wife had not got such an unpronounceable name to pass on. It was Arenwain'on, containing three nasalized vowels and the accent on the final syllable. But I was left in no doubt that it was really Biorn who was being honoured and that we were merely being accepted along with him. Each man who came from the village brought news of the child. In the men's circle they gave graphic imitations of how fast Sibupa could crawl; how he could sidle up on all fours and steal things while nobody was looking; how Pia bathed him; how he cried. I was able to reconstruct most of his daily activities from their reports. One day I was told that Sibupa had fallen into the river—like that, with his eyes open. Arenwain'on was washing clothes. She cried out and a young Shavante dived in and got him. And then there was the day on which a messenger reported that Sibupa could run.

'Standing up, like this?' I queried unbelievingly.

'Yes, like this.'

He was most emphatic about it. Biorn could walk. Not that this was so surprising. He was fourteen months old now and might have done it a lot sooner had he not been so big, but I felt sad and a little cheated. I remembered the terrific sense of domestic triumph that day in São Paulo when he had stood up unaided for the first time and wished I had been there to see him beginning to walk.

It was not so very long after this piece of news that some children skipped up to me one morning and said: 'There's a man from the hunters here,' meaning somebody from one of the other hunting bands. I found the newcomer in his uncle's shelter, lying flat on his back and staring into space. I lay down beside him and for half an hour we chatted about trivialities, our conversation punctuated with long silences. Then I got up to go. At this he got very agitated and I realized that I had committed a breach of etiquette by taking my leave before he had got to the point. Consequently he phrased his message with unusual bluntness.

'Sibupa is dying,' he blurted out. 'Arenwain'on sent a paper to tell you.'

'Where is this paper?' I asked, feeling suddenly as if I had no blood left in my body at all.

'It is with the others. I did not bring it. I was only sent to tell you.'

I cursed the stupidity of the man. Incidentally, his stupidity cost him his life later. But now I knew that there was nothing for me to do but wait through the long day for him to speak at the evening assembly. Nobody would go with me to fetch the letter before tomorrow morning at the earliest. I simply had to be patient. I can still remember how he stood in the men's circle that night and I can hear his voice saying:

'Sibupa is ill. Sibupa is ill.

'He cries very much. He cries very much.

'He is very hot. He is very hot.

'His mother has much sorrow in her belly for the father . . . ugh!'

The grunt marked the end of the phrase and then the opposing speaker would take him up.

'*Ihe, Ihe*.

'He cries very much. He cries very much. . . .'

I crept away to my shelter and tried to sleep, but the ground felt too hard and I lay awake in the bitter cold, watching the moon that seemed to hang motionless in the sky.

Next day I put my few belongings together and the messenger took me back to his own group. I don't know whether it was because I was worried and, therefore, less intent on the trail we were following or whether it was due to plain stupidity on the part of my guide but the going seemed much worse even than when I had travelled with Pahiriwa. Two incidents particularly stand out in my recollection. Once we were leading the pack-horse through a shallow stream. I was coaxing it from behind and seeing to its foothold, the guide was supposed merely to lead it out by its halter. He scrambled up at a point where the bank was steep and the horse only just managed to follow him after much scrabbling, depositing its entire load in the process. S'rimri looked at the mess and then at me.

'I'm hungry,' he said, rubbing his belly, and wanted to sit down there and then and eat some of the spilled manioc flour.

On another occasion he led the horse into a morass where it began to sink. The terrified animal floundered its way deeper and deeper while I wallowed beside it trying to get its load off. S'rimri stood around for a while and then fetched a stout stick.

'Beat it,' he suggested. 'That will make it get out.'

I had an overpowering desire to beat him instead. We had to unload the animal six times in all, three times when it was already in the water and once when he had got it so tangled up in a thicket that it could not move. My guns fell into the water and the loading rod of the ·22 was lost. I was fed up with S'rimri by the time we reached the other camp.

Here I was left high and dry as before, but this time I had a better idea of what to do. I noticed that the age-set who were segregated in the bachelors' hut were travelling with this band so I moved all my things into their shelter. Then I despatched a little boy to find my 'paper' and bring it to me. He returned with a soiled sheet of exercise paper and a small bundle. The message said that Biorn had been running a high fever for days but that he was over it now and seemed quite cheerful. It also mentioned that Pia had baked me some biscuits which she was sending by the same runner. The boys in the bachelors' hut could not understand why I sat down on the ground and laughed so long and so foolishly. They wanted to investigate the bundle, so I shared out the precious cakes while I read and re-read Pia's lines. I did not at that moment care whether I had any food or not.

The senior man in this particular band was Waarodi, the chief's eldest son, and he seemed uncommonly anxious to get rid of me. Unsubtle as ever, he never lost an opportunity to ask whether I did not miss my wife and child or whether I did not find it too uncomfortable out on trek. As it happened I did miss my wife and child, and when the boys in the bachelors' hut ate up my reserves of food when I was out one day, I determined I would no longer repress my inclination to get back to the post. It was about time for the other bands to be getting back anyway. So the next time a friendly young warrior set out to make the journey back to the village, I went with him. I found him congenial company. Since he was a particularly good shot with my rifle I could tease him and claim that he could not hit a deer even if it came up and licked his hand. In return he ridiculed my all-round abilities as a woodsman. We were in a lighthearted mood when we set off towards the Rio das Mortes, and it was not till some time later that I discovered why Waarodi had so urgently wanted me out of the way.

10

Interlude on the river

When I got back to the post I had no desire to do anything save lie in my hammock and watch Biorn. The water in the creek was low and clear. Pia and I bathed again and again, letting the water dry on us as we talked and talked, savouring the manipulation of our own languages. But the idyll did not last. Biorn came down with fever again.

We told Dona Sarah about it and she brought one of the workmen over to see the child. He studied Biorn minutely and then announced that he had been bewitched.

'In that case it can only be the cowboy who has bewitched him,' Sarah interpolated. We never discovered why he particularly should have been held responsible. I did not know what to make of this sudden reversal of roles for Sarah herself. She acted as nurse at the post and we usually saw her dispensing penicillin injections rather than antidotes to sorcery.

'Let's find out,' said the man, and he went outside to search for something, returning a moment or two later with a few tiny flowers which looked like three-leafed clovers. He dismembered them carefully and laid leaves on the baby's head, breast and in his outstretched hands. Biorn clutched the leaves in his little fists as he panted with the heat that was on him. The diviner bent over him and looked long and hard. Slowly he raised his hand and I met Pia's eye. What was he going to do and ought we to allow it? But he just dangled the limp hand over the baby's face and then swung it slowly backwards and forwards like a pendulum. Back and forth, back and forth it went. I could detect the sweat trickling around my ears, and still he went on swinging his hand for what seemed like an age but was probably only a few minutes. Then he gently unclenched the child's fists and took out the leaves. They were not withered. He straightened up as if a great load had been taken from his mind.

'It's all right,' he said joyfully. 'He has not been bewitched, no. He's sure to be all right.'

We did not share his optimism. Admittedly Biorn seemed

perfectly cheerful the following day but he was still running a temperature and we were frightened that he might have contracted some obscure tropical disease which would plague him for the rest of his life. It was more than ever imperative that he be taken to the coast. The problem was—how? The quickest way to get to somewhere which had air connections with the outside world was on horseback, but it would be a strenuous journey for Pia and might involve swimming two rivers unless the water had gone down considerably. It seemed foolhardy to take Biorn on such a venture. The alternative was to go by boat to Shavantina—a long way and against the stream. Understandably, no one was anxious to set out on such a journey, which would take about a week each way and would be exceedingly hard work.

Then, incredibly, we heard the noise of a motor-boat. Inexplicably, it was coming from the direction of Shavantina. I could not think of anybody who would come right down the Rio das Mortes to where we were—about 130 or 140 miles as the river winds—nor could I imagine where they might be headed for. As far as I knew, there was about as far again to where the Rio das Mortes flowed into the Araguaia and absolutely nothing along its banks, no post, no mission, no backwoodsmen, nothing but the possibility of hostile Shavante. Yet I could quite clearly hear the throb of an approaching motor-boat and it was definitely coming from Shavantina.

'If only it had been going the other way,' I thought bitterly. In two days Pia could have been in Shavantina and she would only have had to wait for the next Air Force plane to take her to the coast.

Everybody was down at the river to see the boat come in. It slewed tantalizingly around with the current as if it might go past without stopping, and then beached at what served for our landing place. A heavily built young man jumped ashore. He had blue eyes and a centre parting to his fair hair. He was also wearing shorts and tennis shoes. We did not need to ask to discover that he was the American missionary from Ariões. But he brought some startling news. He was closing down his mission and selling his boat. Ironically enough he was selling it to the Catholic mission in Aragarças, only about 85 miles away from Shavantina as the crow flies. But to get it there he had to sail it right down the Rio das Mortes and right up the Araguaia, a distance of over 600 miles.

He mentioned to us that he was looking for somebody to cook

for them on the voyage and the obvious idea came to both of us simultaneously. The Air Force plane which flew southwards from Shavantina every week made its first stop at Aragarças. Pia could just as well catch it there. Why didn't she then take the opportunity of going down to Aragarças by boat? It would solve all our problems.

After a hurried consultation we put the plan to him. He said he would be delighted to take Pia and the baby but that he only had four other men on board, all members of his mission or connected with his church, and he felt that he could not take an unchaperoned woman passenger. We were taken aback. Women and their babies travel up and down the rivers of Central Brazil on public motor-boats so that it seemed strange that a party of missionaries should be shy about taking one. Still, they clearly found the prospect embarrassing and would not agree to it. Furthermore, they were getting ready to leave. We had to make a quick decision. I hated to leave the Shavante now that I was just getting to know them—but I would have hated even more to have Pia and Biorn stay and for something to happen to the child. I knew there was no way out of the dilemma. We had spent the last few days trying to find one. So I gave in and agreed to go with them. The missionaries were in a great hurry, but we persuaded them to wait an hour while I got our hammocks and some bags packed. Mid afternoon found us sitting startled and out of breath in the motor-boat, feeling as if we had simply got up and walked out of our own lives.

It would take us at least a week to reach Aragarças, they calculated. After the strenuous life of the past months I did not mind having a week off, but I was not sure that I wanted to spend it cooped up in a motor-boat. It would not be easy to keep Biorn healthy and happy under such circumstances either. Finally, I did not know how I would get back from Aragarças to the Shavante and how long this might take me. The more I thought about it, the more I resented the missionaries their modesty. But it was undeniably pleasant to be carried so easily down the Rio das Mortes. For the first time we understood how swift its current really was and revelled in the sensation of travelling without weariness or horses.

The nights were the worst. There was barely room for all our hammocks in the little boat and we could not put into the bank and sleep out, for the river ran between steep walls of tangled jungle which swarmed with mosquitoes. Besides, we were in the heart of Shavante country and it was by no means certain that the Shavante in these parts were Shavante whom Pia and I knew; or that any Shavante who resented our presence would stop to find out who we were for that matter. We had one Shavante on board with us—a good-natured oaf of a youth from the mission who even spoke a little Portuguese. He was more insistent than anybody else that here we should sleep in the middle of the river.

'Shavante here bad,' he kept repeating, and I remembered how nervous he had been at our agency when Apewen had come down to the river to see what was going on. So we dropped anchor in mid river, closed our shutters against the mosquitoes and lay confined in the stuffy darkness. We were not tired enough to sleep and Biorn was usually restless and irritable. Beneath us, live turtles rumbled like ballast in the bottom of the boat. They can be caught and stored, helpless on their backs until such time as one needs to eat them. It was usually well into the night before we could get to sleep, and very soon afterwards the early morning cold would wake us again. We were always glad when we could start off again the next day.

Now we were passing the relics of earlier attempts to contact the Shavante. A small hut, grass grown and disembowelled, marked the site where an expedition had camped and tried to befriend the Indians. Further on a shattered cross was barely discernible at the point where the Salesian fathers Sacilotti and Fuchs went ashore to talk to a band of Shavante. Their companions begged them not to, but they had come to meet the Shavante and meet them they did. They were clubbed to death on the spot. It was some time before their friends dared to return and collect up the bodies, which had literally been battered to pieces, for burial. They erected a cross which the Shavante came later and knocked down. It was re-erected once, but the Shavante threw it down again so the local travellers gave up the struggle.

So we proceeded warily down the river whose name is Many Deaths and were doubly glad when we sighted the great, slow surge of the Araguaia ahead of us. It meant that we were nearing São Felix, a township where there might be fruit and eggs for sale, and that we would soon be able to sleep out on the river beaches.

São Felix, though, was a disappointment. We had looked forward eagerly to our arrival there, for it was, after all, on the Araguaia River, an artery of communication as important as the Tocantins itself. But we found it to be a dispirited place with small, stuffy shops full of canned goods which nobody either wanted or could afford. There were no oranges, no lemons, no chickens, no eggs. The only thing to be had in abundance was liquor, and knots of Karaja Indians shuffled through the sandy streets thinking up ways and means of getting enough money to buy it.

The Karaja are a river-going people who use the Araguaia as their main street. They are famous throughout Brazil for their sculptured figurines and for their colourful ceremonies which have become something of a tourist attraction for those who can afford to fly into the heart of the country for Indian dances. The ones we met, however, were not particularly colourful. Like the Sherente they went about in tattered but 'civilized' clothing and appeared to spend their time gossiping in their strange Portuguese about the price of local commodities. On one topic, though, they were in full agreement with the Brazilian inhabitants of São Felix . . . they both hated and feared the Shavante.

Historical records show that the Karaja have fought the Shavante for at least 175 years, which gives them considerable seniority over the Brazilians of São Felix who have not been doing it for much more than a decade. But we had not been in town more than an hour when it was made quite clear to us that both communities went in fear of the savages to the west and south-west. Shavante, they told us, were bloodthirsty barbarians. Their customs were indecent, their food revolting. It was well known that they were sadistically cruel, that they took no prisoners, that they ate their prisoners, that they reared their prisoners as slaves, that they had harems of captured white women. In short, they could not be trusted. They were astounded to think that we should take the trouble to go and stay with such people and invariably horrified when they discovered we actually had one of the monsters along with us. One of the shopkeepers preached him a little sermon.

'You tell your people—not kill Christians!'

'Shavante not kill Christians!' replied the boy stolidly. 'Shavante not savage, no!'

'They kill them, yes. It's said they eat them too.'

'Don't eat them, no. Shavante only eat game.'

'Well you tell the others not to eat any more Christians. . . .'

It was an absurd wrangle, leading nowhere and without possibility of resolution—rather like the Shavante-Brazilian relations in this part of the country, I could not help feeling.

It was good to get away from the claustrophobic atmosphere of São Felix and breathe the free air of the river again.

That night and all the following nights we camped on the sweeping river beaches. It was always a rush to get everything done before nightfall. One minute we seemed to be fighting our way steadily upstream in the heat of the afternoon and the next the sun was dipping towards the horizon and we were frantically looking for a beach. As soon as we grounded a party would take the guns and go off to fetch firewood. This was no quick job, for the beaches were often half a mile wide at this season and there was a long walk before we got to the trees. Occasionally we would discover that our 'beach' was in fact a sandbank, separated by shallow water from the mainland. This was where the guns came in useful, to scare away the alligators before we made our crossing. Meanwhile our captain and his pilot would be cleaning the engine and Pia getting out the kitchen utensils and cooking a meal.

It was the only time of day when Biorn had an opportunity to run about. We were so nervous at the thought of his managing to tumble into the water during each day's journey that we kept him with a cord around his waist, tethered in that part of the boat where our floor space was. So it was like taking a dog off the lead when we reached our camping place in the evening. He literally skipped with joy and as often as not lost his balance in the process, for it was fine, white sand—so dry and clean that it squeaked underfoot as we walked on it. I always returned early from wood gathering to dig our beds. This might not seem a particularly onerous task but anybody who has tried to dig a hole with his hands in fine sand while a fifteen-month-old child, imagining the whole performance is entirely for his benefit, leaps into it at ten-second intervals will know what it entails. I used to dig a shallow 'double bed' in the sand for Pia and myself, taking care to leave two hollows for our hips, since sand makes a surprisingly hard mattress. At the head of this shallow pit I dug a smaller hole about three feet deep for Biorn. It was imperative to get his finished before darkness and the dew overtook us. We wanted him to be able to sleep dry at least. Over his hollow I erected a home-made tent consisting of our plastic sheeting over a frame of sticks. The only exit from his stuffy little shelter led straight on to our faces as we lay with our heads towards it, and in this way we ensured that he did not crawl away from us during the night.

This was an important consideration for the alligators were our constant companions. One had only to leave the circle of the fire and go down to the water's edge to see the cold glint of their eyeballs, rolling like pebbles with the lapping of the river. We could not begrudge them the river. It was their domain and we expected them to be there, just as we expected the notoriously bloodthirsty *piranha* of the Araguaia to be lurking somewhere out there too. We simply took care how we washed our plates and when we bathed we did so as gingerly as was compatible with our self-respect. But the first night we camped out Pia had gone quietly away from the firelight . . . rather too quietly for she surprised an alligator which had come out to snooze on dry land. It streaked into the water as she gave a startled shout, and when I came sprinting (or rather stumbling) through the sand with my torch I could sense rather than see the multiple splashes as a whole undulation of alligators churned into the river.

It was largely for their benefit that we stoked up the two large fires close to where we lay on the sands. Our Shavante was put in charge of them.

'Shavante don't sleep,' the missionary assured us. 'He'll keep watch for us and keep the fires burning. You need not worry. He's still in his own country here.'

After which our host retired to his elaborate tent and we saw no more of him till the following dawn. Biorn used to wake up twice a night on the average, and it was just as well he did. Each time he came scrambling sleepily out of his hole on to us, I would get up and find the fires burned right down. I put on more wood and fanned the blaze while our Shavante look-out mingled his snores with those of the missionary. In the morning we used to examine the tracks of those animals which had come down from the forest to drink during the night. Their distance from our bivouac was invariably in direct proportion to the state of the fire when I had woken up and tended it. There was always deer and tapir and capivara in profusion, interspersed with the occasional, sinister traces of a lone jaguar.

Day after day we chugged southward. We passed beaches where families of Karaja were encamped, and we marvelled at their pyramids of turtles' eggs, at their catches of monster fish and at the alligator skins drying in the sun. We passed an occasional backwoodsman's hut, similarly festooned with fish and alligators, but sometimes with tobacco drying in the sun as well. We never stopped, for our host was in a hurry—yet our progress was painfully slow. We used to ask the pilot how far there was to the next hamlet along the river.

'São José it's called, isn't it? That can't be very far now.'

'Holy virgin!' he would answer, scandalized, 'it's far.'

We discovered that this was his unvarying answer to questions of distance, and we learned to gauge how many more days there were to the next place by the conviction with which he delivered his reply.

When we arrived at one of these little villages, our host assumed a businesslike 'leave it to me' air.

'Now I'll go ashore and get the provisions. The rest of you stay here by the boat. We don't all want to go wandering into town, because somebody's always missing when we want to leave and we lose time that way.'

His Brazilian assistants did not take kindly to this unseemly hurry. What difference did an hour or two make in Mato Grosso, on a journey which was going to take weeks? But they humoured him. They let him get out of sight and then strolled off on their own errands. What bothered us was that his search for provisions was not very successful. We desperately needed something other than beans and meat for Biorn, yet he would come back with some gasoline and perhaps a sack of flour and say,

'Let's go. They didn't have anything else. Where is everybody?'

At our second stop I accompanied our host into the first shop. He entered it briskly and breezily as if it had been a supermarket in the Middle West.

'*Bom dia*. How are you?' he said. 'Have you got any eggs, or oranges or milk which you could sell us?'

'Haven't got any,' answered the startled shopkeeper. 'Anything else?'

A moment later the missionary was out in the street again and striding towards the next booth.

'You'll see,' he said. 'It'll be just like the last place. They don't have anything here.'

I returned hastily to the boat and reported progress to Pia.

'But he'll never get *anything* that way,' she exploded. 'Come on. Why don't we do the shopping?'

She lifted Biorn on to her hip. He was quite a weight now, but I could not, in Central Brazil, carry him and let my wife walk unencumbered. I would have been the laughing stock of the town. We climbed into the single street. The missionary was already on his way back.

'It's no use,' he called out. 'There's nothing to buy here. We may as well get going.'

'Hold the boat for us,' said Pia shortly. 'We'll get something.'

We spent a long time in the first shop. Chairs were offered us and we sat down and asked after the family of the shopkeeper. We heard all their names and those of them who were present were paraded for our inspection. We gave them news from down-river and in exchange we were told of all the local happenings. Little by little Pia steered the conversation round to children and the difficulties of feeding them. She told them of our journey and of how far we had to go. She explained that she was at her wits' end to know what to give Biorn. They suggested that Dona Ana might have some eggs if her chickens were laying. At the next shop a half hour's conversation resulted in the information that Dona Maria had an orange tree and might let us have some fruit. Dona Maria herself told us that somebody's else's sister might have baked some bread which she would let us have for the child. That day we held up the boat for over two hours, but we came back with eggs and oranges and loaves of bread. We also came back feeling very pleased with ourselves.

Our host did once bring off a spectacular coup. We had run out of meat. One of the turtles died and had to be thrown overboard. The other one was all but eaten up. Only the tougher parts were left—and the soft parts of a turtle take about an hour to chew. We had seen no game for two days. So we decided to break our own rule of 'no stopping en route' and call at the house of a backwoodsman whom we hoped might have a pig for sale.

The missionary jumped out of the boat and disappeared along the path leading to the hut with the gait of a man leading his troops in a final charge. He took the woman of the house completely by surprise. Before she could launch out into the long story about how her husband was away in the fields and would not be back before nightfall and she didn't like to do anything without consulting him, our host had talked her into selling a dozen sweet manioc plants. He took them out of the ground himself, beneath her startled gaze, selected a piglet from her husband's pen, caught it, fixed its price, cut its throat and started back to the boat. As we walked back towards the river with the sucking pig still twitching in his hands, I could hear the poor woman still saying helplessly that she was not at all sure that her husband really wanted to sell, that the price of pigs was not very high at the moment. . . .

That night, with the prospect of roast sucking pig for supper, we camped earlier than usual. By now we were getting to know each other's foibles. The fat man of the company always exercised his ingenuity to find himself a sinecure. If it was getting dark before we tied up, he would offer to hold a torch for Pia so that she could see to make the supper. If not, then he would volunteer to mind the boat while the others went for firewood. He was usually despatched to get wood in the end, and one of my clearest memories of that trip is his retreating figure, round and glowing with indignation, disappearing slowly into the woods after the others.

Another of our travelling companions was quite the opposite. When we had eaten it was invariably he that got up from the otiose fireside and went clattering down to the river to wash the skillets. He volunteered so quietly and so promptly for every task that he made the rest of us feel perpetually guilty.

We even had our storyteller to complete the party. He was what one might call a successful conversationalist. When we sat around the fire at night, brewing a hot drink and reluctant to creep into our dew-sodden blankets, he entertained us; and in entertaining us he sooner or later captured the conversation entirely. But we enjoyed it. We used to sit there while Biorn fell asleep in our laps and listen to his hunting stories from the backwoods. They were utterly fantastic tales, yet they sounded completely plausible under the bizarre circumstances in which they were told. Night after night, as the fire burned low and the grunt and plop of the river animals provided a background for his monologues, we heard about jaguars and snakes, about the water monsters that preyed on unwary travellers and about faithful hunting dogs who died for their masters. Once, in a surge of cynicism, I whispered to Pia that we would soon hear the tale of the man who fought a jaguar with his bare hands. We did. But he reserved it for the night before we reached our destination.

'And he sat there, knowing that he only had to take his eyes off the beast and he was a dead man. He could feel the blood running out of him where it had clawed him and he had to keep pushing his scalp out of his eye where the devil had torn it loose for him. But the jaguar was tired too. It just sat there and looked at him, switching his tail. And the dog sat and watched the jaguar. Every time the beast moved the dog growled at him, and every time the poor fellow tried to get to his knife the jaguar got ready to spring. That was the way they found them, just as the sun was going down. You can still see him around if you go down south. In some place like Fazenda or Itiquiro you'll meet him. You can't mistake him. He's still got the holes in his head and the scoops out of his back and arms where the jaguar got him. Still, it's not every man who wrestles with an *onça* and lives to tell the tale.'

We had been travelling for ten days by this time and we were beginning to be seriously worried about Biorn. Try as we might, we could not keep him warm enough during the long cold nights on the beach, and he soon had a streaming cold. We too had colds that were approaching the proportions of influenza. But there was nothing we could do. We huddled together under the heavy fall of dew and wished each morning that the sun would hurry up. Worse still, Biorn got diarrhœa. That's what we called it anyway. In fact, we suspected it was dysentery but we did not dare to admit it, even to ourselves. Pia especially would not entertain the idea, but I think this was out of consideration for my feelings. In the flurry of our departure I had seized one of our two tins of medicines—the one which contained such useful items as powder for itching feet—and left behind the one which contained our sulpha compound for the relief of dysenterous symptoms.

It was touch and go as to whether we would get into Aragarças the next night. We might if we were lucky, but it would probably involve pulling the boat through the rapids after dark. Alternatively, we could camp out for one more night and try it in the morning. But by this time all of us were impatient. For all Pia and I knew there might be a plane leaving for São Paulo the very next day, and she could catch it if we hurried. We voted to continue.

In mid afternoon we ran aground. It took us only a few minutes heaving to push the boat off again, but from then on the going got more and more awkward. Just before dusk we reached the first series of rapids. There was no waterfall, just an imperceptible incline nobbly with rocks and the water spurting at us as if it was coming out of a tap. The captain ordered everybody but the pilot out of the boat while the two of them took it through the passage. It was a remarkable sight to see the unwieldy craft shuddering with its engine full on, yet completely halted by the water. And yet not completely. Inch by inch it made its way through the archipelago and then heaved into the calm reaches beyond.

The next passage was more difficult. We tried the same technique, but this time the boat stopped and strained against the water like two contestants in a tug-of-war. For minutes the machine and the river wrestled and then the machine lost. The boat began to slip back and slew round. The missionary yelled frantically to the men on shore and we grabbed the tow ropes to bring the boat in alongside the bank before it was swept downstream. There was nothing for it. We would have to heave it over the rapids just like the early explorers used to do. Their boats, of course, were much lighter—but then they had no engines to help them.

It was not a simple operation and required careful planning. Pia and Biorn were directed to stay in the boat. The pilot took the tiller, and the storyteller explained that he was burning with desire to join in the tugging but he had hurt his foot and, therefore, regrettably, could not help us. The fat man's protestations had been forestalled. We needed his weight and he was directed to anchor one of the ropes before he even had an opportunity to suggest he might hold our coats for us. All of us, led by the missionary, stripped off and prepared to heave. The boat nudged its ungainly way out into the racing stream, looking bloated and hydrogenous in the treacherous yellow light of gathering dusk. It took up the slack of the ropes and turned once more towards the rapids, rather like a bull turns unsteadily on the matador. I could see the pale splodge of Pia's face at the side and then there were cries of alarm from the storyteller who had appointed himself foreman and director of operations. A moment later we were all straining at the ropes till we felt out bodies would burst.

We dragged our craft into the rapids and then, the first force of that cataclysmic heave being spent, found ourselves jerked into the water like startled fish. The missionary was shouting at the top of his voice and heaving like a madman. I was up to my armpits in water. The wet rope burned my hands and I could not get a foothold in the rocks. I had a sensation of tearing and cutting at the soles of my feet and remember thinking that at least if I was cut by rocks I was not going to get speared by a sting ray lurking in the sands. We slithered and threshed around the craft and I caught a glimpse of the Shavante, grinning all over with excitement as he struggled on the rope. We were holding the boat now, but not gaining anything. I began to feel terribly cold.

It was probably worse for Pia, however. She told me afterwards that she never thought we would get the boat through. At one moment the engine was straining so that the whole thing throbbed with effort but we were still unable to help it through the rapids. Instead it seemed as if the boat would drag all its boat-men with it. It began to heel over slightly towards the shore and she began to wonder whether it would be better to take the baby and jump for it or to risk being swept downstream. Since both alternatives were highly unattractive she hesitated long enough for the dilemma to be resolved for her.

Suddenly we were winning. The anchor men on the ropes could emerge from the water and step carefully along the bank. The rest of us heaved and waded, waded and heaved until that wonderful moment when the ropes went slack in our hands and the engine came into its own again.

Pia had hot cocoa made for us all by the time we clambered back into the boat again. But all the cocoa in the world could not stop my teeth chattering, and I arrived in Aragarças, our journey's end, with a high temperature.

It was this circumstance, together with my guilty impatience to get back to the Shavante as quickly as possible, which must have made me so obtuse as regards little Biorn. I realized he was ill, but then everybody in the backwoods is ill in some way or other, and I thought that a visit to the doctor as soon as he got back to São Paulo would probably set him to rights again. My immediate concern was to get back to the Indians and to get myself well enough to survive with them.

As luck would have it, the same special duty Beechcraft which had taken us to the Shavante was at Aragarças. The commander at once agreed to fly me back to the Indians, and I left rejoicing, firmly convinced that Pia and Biorn would soon be on a joy ride to the coast. But they were not so fortunate. Their plane to São Paulo was days late in coming, and meanwhile Biorn got steadily worse. His 'upset stomach' was clearly developing into a serious condition. By the time the plane arrived he could barely hold his food and, worst of all, he did not cry any more. He lay and looked at his mother like a sick animal. She was in despair, especially since it was by no means certain that the Air Force plane would be able to take them. It did, of course. As a rule Brazilian Air Force officers are both helpful and gallant, and here was a case where they could combine both virtues.

In São Paulo Pia made the taxi take her to a hotel where she knew she could be sure of having a telephone in her room. If she was going to be confined with a sick child she wanted to have some means of getting in touch with her friends. But the recep-tionist cast a cold glance at her expedition clothes, at the ailing baby and at the Shavante baskets which held much of their baggage. He decided that there was no room available. She explained, I gather, pretty forcibly who she was and what she had been doing and was straightway ushered into a room with a telephone.

It was exceedingly fortunate that she insisted, for the first thing she did was to call a good friend in São Paulo and ask her to come round. She did, took one look at Biorn and called one of the best pediatricians in the city. He came and ordered Biorn straight into hospital. Pia did not even have time to get a meal. A few hours after her arrival she found herself, haggard and weary, keeping watch by Biorn's bedside while he was given blood transfusions and a saline drip. They told her that he had dysentery and that he was completely dehydrated.

11

A strong chief

Mine too was a cheerless homecoming. The Beechcraft did not even stop its propellers. I got a handshake from the fliers and they were off almost before my feet touched the ground. Nobody came down to the airstrip to see who had arrived. I trudged into the village and made my way to my hut with the usual dogs barking around my ankles. The Shavante were quick to notice my feverish cold and it made them jittery. I had been amongst the Karaja, they said, and the Karaja were great sorcerers. Clearly I was bewitched. In any case, they avoided me.

I lay in my hammock letting the fever run its course and watching the lizards as they gradually lost their shyness of me. I must have stayed there for a couple of days or more. When I roused myself at last I remembered that Pia had promised to contact a radio amateur in São Paulo and send me a message. But I no longer knew what day it was! Then I remembered that the elders were keeping track of the initiation ceremonies by making a notch for every day in a ceremonial pole erected at one end of the village. On the basis of their calculations I discovered that I should hear from São Paulo the very next morning.

We had arranged to make contact early when reception was at its best, so I made my way up to the post in the cold half-light before dawn. The metal containers of the radio transmitter were battened to the single bench by a complex system of cobwebs. Mice bolted as I came in and spiders scattered. I checked the corners for snakes and then set up the batteries. Two Shavante appeared in the doorway and goggled at my headphones.

'What to do?' they asked.

'Speak to Arenwain'on,' I replied with more conviction than I felt.

Amateurishly I adjusted and tuned. There was a lot of static. The Shavante kept up a fire of questions which merely increased my nervousness.

'How will you speak to her?'

'Where is she?'

215

'Will she speak to you?'

'Is that her?'

It was the pilot of a Pan American plane talking to his base.

'No, no. Hadu. Don't talk so much.' I was getting irritable as the time approached. Was my watch wrong? Perhaps she had already spoken. Perhaps I was not tuned correctly. Perhaps she had never got in touch with the radio operator in São Paulo. Perhaps . . . perhaps. There were so many possibilities that it became of overwhelming importance that I should hear her, simply so that all these uncertainties could be cleared up.

I could hear a number of things but not Pia. I tried calling her up myself, but without result. After fifteen minutes I was about ready to give up when I heard, loud and clear, my own call sign and a man's voice announcing that this was such and such an operator in São Paulo co-operating with the expedition to the Shavante. I was so eager to switch over and start talking to him that at first I did not listen very carefully to what he was saying. Then it was brought home to me that it was something about a hospital and I began to listen very carefully indeed.

'. . . but she will be able to speak to you tomorrow at the same time. Over.'

I begged him to repeat the message and asked him if he was receiving me, but all to no purpose. The tenuous link between us seemed to have snapped. All I heard was static and the slightest flicker of my dial brought me dance music or military aircraft so that I was afraid even to search for my interlocutor. A moment later he was there again, telling me that my wife and son were in hospital, though out of danger, that she hoped to be with him the following day to speak to me in person and was I receiving him?

It must have been about 7.30 a.m. when he went off the air and I was left wondering how I would get through that long and tedious day, waiting to hear more news. I checked everything that I could think of on my radio, my aerial and my batteries to make sure that they were in as good order as possible for the next morning, then I went back to the initiation ceremonies.

The following day a troop of Shavante came with me to the post to watch me talk to Arenwain'on on the talking-thing. They sat about, surreptitiously fiddling with the dials of the radio and then loudly upbraiding the children for doing exactly the same thing. I was afraid that someone was going to get an electric shock, and even more nervous at the thought that through some clumsiness on my part I might not hear her. If only I could get

Shavante: Waarodi (*facing*) and Rintimpse making simultaneous speeches

Shavante mother with her baby (*left*)

Shavante: Hearing their songs played back (*above*)

Shavante: Chief and his sons
out hunting (*above*)

Biorn sleeps out by the Araguaia River (*left*)

through to her today, then we could arrange other times at which to talk and I would have a link with the outside world at last.

I found them fairly easily this time. First the radio operator and then Pia came on the air to tell me that Biorn was still in hospital. She did not know what to do. They had planned to return to the Shavante after a short while in São Paulo, but this was clearly out of the question. What did I suggest? But they could not hear me. I could even hear them in the room together, discussing how they might try and reach me and finally agreeing that it was no use. Their voices were getting fainter and fainter and I realized that my batteries were too low. Those months in the customs sheds at Rio had affected the accumulators and I had had no opportunity to recharge them before going into the interior. So now there was nothing I could do but sit helplessly by and listen until they got bored with talking into the silent air and switched off.

I cursed the Brazilian customs, and the radio and all the time and trouble and anxiety it had caused us. Now when we really needed it, it let us down. I disconnected it savagely and went back to the village, consoling myself with the thought that in spite of all these setbacks I was making a little headway with the Shavante.

I decided to visit S'rimri and see his new baby, born two days after he had come to me in the Roncador with the message that Biorn was dying. I had not seen him about lately, I thought as I ducked into his hut. I stood for a moment to let my eyes adjust to the darkness and then I got a shock. S'rimri's wife who had been perhaps the most beautiful girl in the village was sitting there with her head completely shorn. There was snuffling in the darkness as I approached and I could just make out her mother sitting there with her head shaved too. They were both in mourning.

'Where's S'rimri?' I asked.

'He's dead.' It was the mother who spoke. The girl sat staring into space and rocking her baby. Her mother, who used to give me all the village gossip, now had tears running down her cheeks, and Indians do not cry easily. 'He's dead,' she kept repeating. 'He's dead. We miss him a lot. He was a good man and now he's dead.'

'But how did he die?'

'They killed him . . .' The story came out in a flood of words and sobs. 'They' had apparently seized him while he was out

hunting and held him while others fired arrows at him. Then
they had clubbed him till there was no life left in his body.

'But who? Why?' I pressed her.

'Those ones from over there. . . .' She lowered her voice.

'What are their names?'

She told me.

I was astounded. The three men she named were three of the
warriors with whom I was most friendly. One of them called me
'brother' and we had hunted together, even slept night after night
under the same blanket. The others were those of the young men's
age-set whom I would have thought least likely to commit murder.
Yet S'rimri was clearly dead and why should the murdered man's
mother-in-law, in the midst of her mourning, invent such a circumstantial story about the events of his death?

On and on she talked, pouring out all her troubles to me. Her
daughter, as impassive as ever, went to the centre of the hut and
busied herself with the cooking pots. I lay on the dead man's
sleeping mat and let the hysterical flow of Shavante wash over me
as I racked my brains. There must be a reason for this murder.
Yet I, who was beginning to think that I was making progress in
understanding the Shavante, neither knew that murder had been
done nor why it had been done.

It was unlikely to be an inter-clan dispute since the murderers
were from different and mutually hostile clans. I understood only
one thing, and that was why Waarodi had been so anxious that
I should leave his band when we were out on trek. S'rimri must
have been killed while I was still on my way back to the post.

I went over to spend the heat of the day in Apewen's hut. There
were usually plenty of people gathered there to gossip, and I
thought I might learn something of S'rimri's fate. Instead I
suffered my second mortification of the day. Apewen's daughter
was dandling a tiny baby. As my 'sister' she had looked after me
while we were on trek. We had lived in the same tiny shelter and
spent many a night in the pile of bodies that competed for the
warmth of my blankets. Yet I had not realized that she was
expecting a baby. It may be said in my defence that Shavante
women of a certain age, and that means in their twenties, develop
very pronounced bellies, so that pregnancy is not so noticeable,
even though they do go naked. Besides they are so active. I
reflected that it was probably the fact that Wauto would go out
collecting and come staggering home under awesomely heavy

loads which had blinded me to her condition. Still, it was another blow to my self-esteem as a keen observer.

It was small comfort that the mature men called to me that night as they went to their evening council or that a number of them patted the ground invitingly beside them when I joined the circle. I was beginning to be appalled by my own ignorance of them and at the enormity of the task which lay ahead of me if I hoped to gain even a limited understanding of their society.

Apewen rose to his feet and began speaking. He delivered his remarks in the formal, rhetorical style which Shavante, just like their cousins the Sherente, consider of such importance. Each phrase was delivered in a low voice, fast and unaccentuated, rising to a final explosive syllable. Then the phrase was repeated and the next one came tumbling on its heels with the controlled breathlessness of oratory. Before long the senior man of another clan had also risen to his feet. He took up the chief's points antiphonically, repeating them and rephrasing them at the same time as the old man was talking so that the whole performance sounded like a symphony of words. When I first came among the Shavante I had been fascinated by the sound of their debates. Tonight I realized that that was all I could do and all I was likely to be able to do for a long while to come. Later I felt it a minor triumph when I could follow the gist of what the speaker said.

Nobody could help me learn the language. The Shavante knew no other language into which they could translate and, what was worse, they were so unused to dealing with outsiders that they were incapable of putting one idea into different words for the benefit of foreigners. If I lost the thread of the speeches and nudged Urbepte to ask what was being said, he usually replied, 'He is very angry,' or, 'He talks very much.' Here was all this priceless information being flaunted before my nose every night and I was incapable of taking advantage of it. It was like taking an examination in unseen translation from a language one only knows poorly, failing it and being faced with the certainty of having to take a similar one every day in the foreseeable future without ever being permitted to consult a grammar or a dictionary.

I lay awake for a long time that night, wondering. As soon as I had gone to sleep I was awakened again by a sharp, rustling crash as somebody came through my thatched entry.

'Apewen,' in a hoarse whisper, 'Apewen, come out and sing.' It was the initiates who were going on a singing tour of the village.

'Sing, nothing,' I said rudely, 'I am sleepy.' A few minutes later, as it seemed, someone was shaking my hammock. This time it was Suwapte.

'Brother-in-law,' he said, 'come out and sing.'

Outside I could hear the deep baying of the flutes and the call of 'my' age-set, the *Ai'riri*. We were sponsoring the initiation festival, and part of our obligation was, therefore, to dance before dawn and before dusk each day. It was bitterly cold, and I could feel the mist rather than see it when I stepped outside the hut. The *Ai'riri* already had a good fire going in the middle of the village and I made my way gratefully over to it. Then the master of ceremonies put away his flute and we all joined hands in a circular dance.

Usually I quite enjoyed the dance and the singing that went with it, but today I was feeling introverted and tired. I kept thinking that a Shavante, under similar circumstances, would have lain down on his sleeping mat and refused to speak or be spoken to until he felt sociable again. Why, I reflected bitterly, did I feel I had to be everybody's good-natured fool, just because I was an anthropologist? Surely there was a better, a more sophisticated way of understanding the Shavante than by dancing around in the cold? But then ratiocination had not got me very far either.

For the moment at least I had more than enough to do simply keeping track of the ceremonies. The boys in the bachelors' hut were being initiated into manhood; but there was no isolated ritual which effected this transformation. About the time of our arrival they had had their ears pierced to accommodate the three-inch cylinders which Shavante men always wear in their lobes. The symbolism of this is quite explicit. Shavante believe that a man has only to redden his ear-plugs with life-giving urucu and wear them while he has intercourse with a woman for her to conceive. Similarly the cylinder pierces the lobe of a grown man's ear just as he may now enter a woman. The boys, then, were adjudged to be sexually mature, but they were still not quite men and still not permitted to exercise their sexual powers. Instead they were obliged to rise before dawn every day to run a series of races interspersed with ceremonial parades around the village. Old Apewen was invariably out there in the cold, keeping them at it, and the mature men would gather round his fire towards day-break and shout encouraging remarks to the tiring runners. The initiates had to run another series of races at dusk and then, when they had flopped down exhausted in their hut, they were usually fetched out during the night by one of the members of my very own age-set, for we were sponsoring their initiation, and obliged to go singing round the village. In this way they acted out those masculine virtues which the Shavante most highly prize: wakefulness and fleetness of foot.

There was still another important quality which they had not been called upon to demonstrate and that was hunting skill. So I was not surprised when I learned that the mature men were to sponsor a lengthy hunt for them which would keep us all away from the village for a week or more. It was with many misgivings that I accompanied them, and I nearly developed a crick in my neck during the days that followed since I was perpetually craning over my shoulder to make sure that there was no plane hovering over the post. Not that I would have known what to do if there had been. We were some days' journey away by then, and I could never have found my own way back in a hurry. When we returned from the hunt there was still no news from Pia.

In spite of all my anticipation, the plane surprised us. It was through the hoarse singing of the initiates that I heard the crescendo of its engines as it came to a stop on the distant runway. It was still there, with Pia standing beside it, when all of us arrived panting.

'I can't stay,' were her first words, 'but I've brought you some food.'

My heart sank.

'You mean you're going back at once? What's happened? Where's Biorn?'

Her answer was drowned out by the roar of the engines. Sand whipped into our faces. 'I'll explain it all later,' I could hear her say.

It was certainly a fantastic story. She told me how Biorn had been rushed to hospital; how they had given him blood transfusions and tried to fix him up with a saline drip. He had struggled and screamed so frantically that they could not get the needle in. Finally they had been forced to strap him to his cot so that the drip could be put into operation. All that night and the following day she had sat with him. Her friends tried to trace her, and one of them even succeeded in getting a call through to her, but when she went into the corridor to take the telephone Biorn woke up and struggled so furiously that he jerked the needle out of his leg. After that she would not leave him on any pretext whatsoever. Nobody in the hospital thought of bringing her any food.

At last her friends had found her and fed her and insisted she rest while they took turns at watching the baby. It was a week before they could leave hospital, and then she was faced with the problem of what to do next. Clearly Biorn could not go back to the Shavante as he was, but we had no alternative arrangements and, worse still, no funds to make any. Once again friends helped out. They looked after Biorn while Pia went down to Rio to see if she could collect the payments on my Brazilian Government fellowship and perhaps arrange for an eventual return to the interior.

But Rio has a reputation throughout Brazil for bureaucratic indifference. Disasters there become dossiers and dossiers tend to get shelved. It was regretted that the money owing to me could not be paid. The requisite papers had not been made out to authorize the issuance of a cheque, and in any case such a cheque could only be issued to me personally.

In desperation she went to the British Embassy. She explained the tribulations of our research, the baby's illness and the technical difficulties in the way of collecting our money. In short, she asked if they could do anything to help. She was received by the same senior official who had previously taken such a keen interest in my credentials, though not in my research. He went so far as to postpone his morning drink in order to explain to her that the Embassy did not like to get involved in such matters. He realized, he assured Pia, how frustrating and well, er . . . awkward such things could be, but really, he was sure she would understand. . . .

'My advice to you,' he told her affably as he made ready for that impatient cocktail, 'is to go home and write a book about it.'

Pia went to the Air Ministry instead where a bland junior officer assured her that there was no airstrip for any post called Pimental Barbosa.

'Pimental Barbosa is the name of the Indian Post,' she explained. 'The name of the place is São Domingos. You used to have a plane that went up there with supplies. The same one that flew on to the Xingu River.'

No. He insisted politely but firmly that there was no such landing-strip, or that if there was it had never been regularly served by Air Force planes.

Pia then became the raging mother from the Rio das Mortes.

There had been a supply plane and it had been discontinued without a word of warning, she told him. Not only had her baby fallen ill and nearly died but there were scores of Brazilian personnel all over the backwoods still hopefully watching the skies for supplies which never came. Did he not feel any responsibility to them?

The young man got nervous and took her in to see the brigadier himself. The brigadier, being a diplomat, ordered coffee and suggested that Pia should tell him all her troubles. She did. She showed him on his own wall map where São Domingos was and pointed out that there was still a hole there where a flag, representing a staging post for one of the Air Force supply planes, had been taken out.

The brigadier explained in a kindly fashion that the capacity of the Brazilian Air Force had been strained to its limits by the tremendous air lift required for the building of Brasilia, the new capital in the interior. Every available plane had, therefore, been diverted for this task. The supply plane for the Araguaia-Xingu area was regarded as expendable when compared with the urgency in this other sphere. Nevertheless, the Brazilian Air Force always had been and always would be willing to do everything it could to facilitate the work of people who were engaged on serious projects in the interior.

'It is sometimes hard for us to tell, *minha senhora*,' he shrugged. 'So many people want to be flown into the interior, just for the romantic thrill of it, and most of them call themselves anthropologists if they cannot think of any other justification for their trip.'

Pia was disarmed by his charm and his candour. The outcome of their talk was that he offered her not only a passage back to Shavantina ('You clearly need to go and talk things over with your husband') but also a passage back to São Paulo as an additional courtesy.

There her friends stepped in again. They paid her bills and minded her baby, insisting that she should implement the brigadier's authorization to travel to Shavantina so that we could make fresh plans for the remainder of our time in the field. From Shavantina she had been brought back to us by none other than Chico Doido (Crazy Chico) himself. Chico was the biggest daredevil among the pilots of Central Brazil and flying with him was a unique experience. On this trip he had been busy scaring deer, spotting them as he flew along the river and then swooping down and sending them galloping crazily into the thickets before he tilted the plane's nose over the tops of the oncoming trees and soared into the sky again, leaving Pia's stomach a few hundred feet below. No wonder she had been somewhat incoherent when she arrived.

I lay back in my hammock and considered the implications of her story. Then a thought struck me.

'How on earth are you going to get away from here?' I asked. 'Did you ask Chico to come back for you?'

She shuddered at the thought. No, she had concocted an even more bizarre withdrawal system for herself. When she went to see the Air Force commander in São Paulo with her authorization to proceed to Mato Grosso, he had put her in touch with a certain lieutenant who was known throughout Central Brazil simply as Tenente Haroldo. Tenente Haroldo liked flying up to Mato Grosso on special missions. In that way he could be on his own for weeks at a time, hedgehopping in the remoter parts of the country, and escape the tedium of Air Force routine down at the coast. He had a foolproof system for getting these assignments which was quite simply that he was one of the few officers in the Brazilian Air Force who would risk his neck in one of their oldest four-seater planes. He had developed an affection for the old single-engined machine, and I suspect that but for him it would have been scrapped years previously. But Tenente Haroldo had a sergeant who understood the flying crate and nursed it through mission after mission with the loving care that a man might bestow on an Edwardian car. Haroldo always volunteered to fly it whenever there was a mission of some urgency but little priority to be taken care of in the interior, and he and his anachronistic steed were beloved throughout the backwoods in a way which the Beechcrafts could never be. There was a touch of the personal, of the knight errant in his unpredictable appearances, He was never expected, and he usually came to help out those in distress.

When Pia had explained her need to get to the post and, more complicated still, her need to get away again, the Air Force authorities scratched their heads and decided that this was a case for Tenente Haroldo. He was collecting packages for isolated spots in the interior which the regular planes did not visit, and he could be expected to arrive any day with a folio of errands to run.

'He is bound to be given a long list of things to do which he cannot possibly get through in the time at his disposal,' I said. 'I only hope that coming here is one of the jobs that he manages to squeeze in.'

'Oh, I think he'll come,' said Pia. She had slipped back into the active, optimistic thinking of the cities.

'It'll be awkward if he doesn't,' I replied with the unconscious pessimism of the backwoods.

But he did. Just when I was congratulating myself on having Pia there to help me keep track of all the various aspects of the initiation ceremonies, the strange old kite clattered into the airstrip.

To enter it was for all the world like getting into a car. There were four seats and you slipped into the back ones by tilting the front ones forward. After Pia and her bag were safely on board, Haroldo and his sergeant went through the rigmarole of their flight preparations. Only this time the familiar procedure was interlarded with their private precautions.

'How does she sound?'

'All right.'

'Are the flaps all right?'

'Flaps all right.'

'Shall I open the throttle?'

'Open the throttle.'

'Will she get off the ground?'

'She'll get off the ground.'

It was not a dialogue to inspire confidence, and I could see Pia getting pale in the back seat.

'Cheer up,' I shouted over the noise of the engine. 'Have you ever met a bad pilot in this part of the world yet?'

She shook her head. To the layman the skill of the Brazilians who flew in these parts was astounding.

'And these must be extra good, or they wouldn't fly this crate at all.'

I don't think Pia was particularly reassured.

In the village the Shavante were distributing meat once again, this time in honour of one of the boys from the bachelors' hut who was to assume the name and ceremonial office of Tebe (Fish). He was dressed in one of the long masks which the men had spent so many weeks preparing when they were out on trek and the master of ceremonies was there too, wearing his largest and heaviest grass cape. All night long they paraded round the village, presenting the cape to each house where the boy beat it rhythmically with his wand, and whistled. I was woken again and again by the soft rustle of the cape like a snake in the grass and the mournful whistling mingled eerily with the coughing of a near-by jaguar. In the background the boy's family wailed and wailed to mark his passage to a new status. It was a true witch's night and for once I was not sorry to get up well before dawn and be one of the first at the mature men's fire. The young Tebe was still whistling.

'*Hu*,' said one of the older men, using the marvellously descriptive Shavante word for jaguar. '*Hu*—they have been calling this night.'

'It's too cold even for jaguar,' another one laughed.

'They come here to drink, the country is all dried up.'

'Nonsense. They come to eat the cattle at the post.'

We could still hear them calling. There must have been at least two of them but it was difficult to tell, a jaguar's whiny, throaty noise seems to come from all points of the compass and simultaneously from ten feet and two miles away.

Urbepte came up to me.

'Let's go, Apewen. It's time to dance.'

'I'm too cold,' I protested feebly.

'Then dancing will warm you,' Urbepte replied inexorably, taking my hand in his and drawing me into the circle.

We seemed to be having an orgy of dancing. All through that day we were summoned at intervals to dance. At dusk there was a respite but only long enough to enable the members of my age-set to paint themselves. They reappeared when the moon rose, eerily decorated in skeletal patterns of black stripes. Urbepte had one eye blacked out and the other picked out in scarlet.

'What dance is this?' I asked him.

'It is the dance of the spirits,' he replied.

All night we danced around the watch fires while the women and children huddled against the cold. I did not know it at the time but we were celebrating a wedding. It was the end of the initiation.

Next day the newly initiated young men lay inert in their own private shelter. My age-set gathered at the entrance and immediately a woman came over leading a little girl who cannot have been more than four years old. The mother placed a maize cake on the ground before us and led her daughter into the shelter. There she made the girl lie down on a sleeping mat beside the boy she was to marry. He turned his back on her and covered his face for shame.

More and more women were coming now, bringing the boys their brides. Each one paid a maize cake to us as the sponsoring age-set and then took her charge in to lie beside her husband. Some of the brides were wide-eyed little things about seven years old, others were so small that they were brought on their mothers' hips. Many of them were terrified by the solemn row of inert men, lying there with faces averted in their ceremonial paint. They shrieked and struggled and had to be pushed into the shelter to sit for a moment beside the boy who had been secured for them.

The wedding ceremony closed when each boy had been provided with a bride. Only in this way could older men take more than one wife in accordance with Shavante custom without preventing the younger ones from marrying altogether. It was true that the initiates would have to wait for a while until their infant wives were physically able to consummate their marriages, but each one was at least guaranteed a bride on being initiated. Nor would he have to wait all that long. Shavante girls are usually deflowered by their impatient husbands when they are somewhere between eight and ten years old.

In any case the newlyweds would not live together. The girls return to their parents as if nothing had happened while each newly initiated warrior parades his manliness and his chastity. He might visit his wife secretly by night, but by day he repudiates her. He is a warrior. He does not need women.

It is his in-laws who watch for the opportunity to prick the bubble of his pretensions. When they discover that he is visiting his bride, they erect a small enclosure in the hut where they can have privacy . . . 'so that he may not feel ashamed', as they put it. In fact, the whole purpose of the enclosure is to publicize his shame and make it known to the whole world that he is not what he pretended to be, a dashing young warrior, independent both in body and soul. Instead he is just a young man whom they have caught sleeping with their daughter, who is their in-law and who will soon move into their house where he will be an inferior, someone who listens rather than talks, someone who runs errands rather than gives orders.

No wonder the warriors whom I had talked to all strenuously denied being married. One of them even denied it after I knew he had a child. It was usually the old women, their mothers-in-law, who gave the game away and told me in lurid detail how each young man came to visit his girl, how sluggish he was in the mornings and how he slunk away when he awoke and realized that the family were aware of his presence in their hut.

No sooner had they completed their initiation ceremonies than the men began to get restless. There was talk of all the deer to be had away to the north. I sensed the migratory urge rising in the community and had to decide whether I would go on trek with them again. I badly wanted to visit some other Shavante villages, particularly the 'wild' community about whom I had heard so much. I had met many Shavante who were from there originally and they struck me as being no more ferocious than their fellows. Besides, Pia was negotiating with a São Paulo newspaper to see whether they could not provide a helicopter to take me there in return for the story. It seemed improbable that she would be successful, but if by some lucky fluke she should be able to get hold of one, I did not want to be wandering along the Rio das Mortes chasing deer. So I decided that I would visit the Catholic mission at Sta Terezinha and talk to the Shavante there where I was more accessible. Thus when Apewen's people went north I hired a guide from the post and headed in the opposite direction.

Raimundo was comfortable and that annoyed me. For four hours now I had been watching his effortlessly straight back and the jaunty angle of the hat on his woolly head. He and his mare were a team. They did not stumble. Sometimes she gave a little more than she intended on one foot, but Raimundo gave with her so that there was something fluid and unbreakable about their relationship.

My own poor nag was full of surprises. Anyway, he usually managed to surprise me and throw me off balance, which threw him off balance and disrupted the rhythm of our progress. I was beginning to remember the importance of rhythm on journeys of this sort, but was afraid that the recollection had come too late. Already I was beginning to be stiff in the saddle, and the more I wriggled the harder I made it for my mount.

We were travelling under a fierce sun through a land where the very shadows had evaporated. It was hard to believe that it was the same countryside which we had seen crinkly beneath the water a few short months ago. Even the coarse, indestructible *capim*, the stringy grass of the interior, was withering with fatigue. Trees drooped and shrivelled, and the thickets crackled like charred bones when we made our way through them. Out in the open the air was thick and yellow as melted butter, and even the flowers which occasionally sprinkled the grassland were so unnaturally brilliant that they gave our world the appearance of a moonscape.

The blanket which I had folded lavishly beneath my thighs in the early morning now felt like a strip of dried meat and gave me as much cushioning. The creak of our saddles and the metallic clink-clink of my mess tins against my horse's flanks were beginning to get on my nerves. There was nothing I could do about it. I had already stopped twice and tried tying them on differently. They always worked themselves loose enough to tap, tap, tap until I felt the noise was inside my own head.

Sometime before midday we stopped at the Lagoa da Onça (Lake of the Jaguar) and ate manioc flour fried with bits of dried meat. The water in our washed-out kerosene bottle was the temperature of tea which had stood for a little while. But the Jaguar Lake was a disappointment when we came to try and refill—it was bone dry. I was sorry we had tried to get water there at all. I felt sure that if we had not been unsuccessful I would not immediately have felt so thirsty.

Raimundo reassured me that there were bigger and better lakes further on. But now the sun had passed its zenith and there was no sign of any more water. I brought my horse up level with his to ask him how much further he thought the next lake would be. He guessed that it was about a league and I groaned inwardly, perhaps outwardly too for all I know. A league was nearly four miles, a good hour's journey, and if it was a 'big league' it might take us an hour and a half. There was also that faint hope that it might be a 'small league' requiring less than three-quarters of an hour, but I put it sternly from my mind. There had not so far been any small leagues on this journey and we had plenty more to go. There was some argument amongst the locals as to the exact distance it was overland to Sta Terezinha, but the figure varied around sixteen big leagues, at least sixty miles or more. Unless we wanted to sleep out two nights running we would have to ensure that we broke the back of the journey today and that meant covering more than eight leagues.

It was mid afternoon when we got to Lagoa Grande (Big Lake) and I dismounted as stiffly as an arthritic coming down some steep stairs. An undistinguished clump of trees with brackish vines trailing on them like torn ligaments marked the site of Big Lake and we marched joyfully in to slake our thirst, keeping a watch for snakes as we went. The dried mud crackled like gelatine beneath our feet and once more I had an uncanny feeling as if we were crossing the porous gullet of a volcano. The trees closed in on us but there was still no sign of water. Then we were on Big Lake, tramping across its withered cheek to the few blades of green grass which marked its centre.

Raimundo stood looking down at them for a long moment before he shrugged in resignation.

'I don't think there's water here either,' he said.

'Not enough for the horses,' I replied, 'but perhaps enough for us. Let's dig.'

We dug—to the full extent of my reach (I had longer arms than Raimundo) and the hole produced a cupful of dirty water like the sludge in a cement mixer. It was just enough to wet our lips and to give us a bad taste in our mouths for the rest of the way.

By nightfall two other 'lakes' had let us down. The horses' heads drooped and their spines were mountain ranges. It had been a long time since I had ridden so determinedly all day, and every fibre of my body ached from the waist down. I let Raimundo ride ahead while I did exercises in the saddle, flexing my legs to stave off the cramps and praying that he would not look round and see my undignified antics. In the gathering dusk the jaguars started snarling and moaning in the distance and still we rode on hoping for water.

We gave up and camped soon after dark. By that time I ached more than I would have believed possible without having been maltreated. It seemed to require an inordinate amount of effort to find a couple of trees that were correctly spaced for my hammock. We unsaddled and hobbled the horses and went supperless to bed, for we were too parched to eat anything.

It must have been a hour or two later when I was rudely awakened by the cough of a jaguar which seemed to come from right under my hammock. I had not really slept—I was too sore for that—but in my painful doze I was aware that there were jaguars about. But I also knew that the call of a jaguar is highly deceptive. The animal puts its muzzle to the ground before it roars and often sounds as if it is ten feet away when in fact it is at least a mile off. But there was nothing deceptive about this.

'Hu, hu, hu, hu, HU, HU!' it roared again, giving the sound that whizzing tone, as if someone near by were whirling a rope round and round his head very fast. For the first time I picked up the six shooter in earnest, but it was a limited consolation. I knew very well that I was at a considerable disadvantage. Even experienced hunters do not like to trail jaguars without dogs by night. I got out of my hammock, not because I wanted to go anywhere or do anything—I did not know what to do—but simply because one is powerless in a hammock, and I felt more secure on my own two feet.

'Raimundo,' I called, 'where's the jaguar?'

There was no reply. I strained my eyes to see if Raimundo was still asleep in his hammock, though it seemed unlikely. The conviction grew in me that he was not there at all.

Again the whining, snarling roar, ending in a cough. It sounded as if the jaguar was within springing distance of me. I could not see him and could not even be sure in which direction he was lurking. I had a feeling of being watched. It is the hungry season for jaguar, I was thinking, perhaps it will attack. I had heard that they did not usually attack men. Perhaps it was after our horses. It would not be very amusing if it killed one of them either. A jaguar could kill any number of cattle in a night, just for the sheer delight of it, eating only a little of each kill. What could it not do then to our hobbled horses? I contemplated the unattractive alternatives of a direct attack on me or a massacre of our mounts which would keep us walking for two more days at least through this waterless country.

I waited until I began to shiver with cold and tension but the beast gave no further sign of its presence. I heard the grasses swishing twenty yards away and took a firm hold of my revolver, thinking ruefully about the absence of low cover to rest my elbows on.

'Raimundo?' I hazarded again again.

'*Sim, senhor*,' came his welcome voice, 'that devil's quite near.'

I put the revolver down.

'Did you go out after him?' I asked.

'Like hell! I don't go after *onças*, no sir! Not unless I have trained dogs and then only by day.'

'What were you doing then?'

'I went to get the horses. Don't want them to be killed.'

The jaguar roared again, but this time I could tell it was some way away, if only by comparison with its previous noises.

'Go to the devil!' shouted Raimundo gleefully into the night. 'You won't get a meal here, you rascal.'

The jaguar disdained to reply.

'We might let him eat one of the horses, just to see where he goes to drink afterwards,' I said, and Raimundo laughed so much that the tree his hammock was attached to showered him with a fall of dew.

The culminating irony of our situation was that both of us were sodden with dew when we awoke, still thirsty, before dawn next day.

'Trouble with this country,' Raimundo muttered as he saddled his horse. 'Either too much water or not enough. Oh, it's hardness itself!' Saying which he swung blithely into the saddle and set off, humming under his breath. I clambered rather than vaulted on to mine, and my thighs started to ache almost at once, just out of anticipation. It was not a day to look forward to, and it fulfilled my expectations of it. By midday we still had not found water and so we did not even bother to stop and try to eat. I gave up asking how much further we had to go and tried to make my mind a blank. I only succeeded in eliminating every thought except 'I'm hungry', 'I'm even more thirsty' and 'I wish this journey would end'.

It did, quite suddenly, in the early afternoon. I noticed we were heading towards thick jungle and called out that there must be water over there. Raimundo laughed.

'That's the river,' he said. 'We are nearly at the mission.'

That meant we must have ridden about forty miles the first day, so I was not surprised that I felt so stiff. We had to cross a stream before we reached the mission. I don't know whether it was through some puritanical desire for self-punishment or just a wish to get the journey over, but I did not want to drink at it. It had been a waterless trip, so let it remain that way. I would drink when I was dismounted, the horses were unsaddled and I could take my ease. But the horses did not see it that way. They drank noisily and lengthily through their bits. Raimundo slipped into the stream and started scooping water into his mouth. I sat slumped on my horse, childishly obstinate and longing for a drink. Finally, just as Raimundo was getting back into the saddle, my sense of proportion returned and with it my appreciation of the ridiculous. I landed with a splash in the shallow water and began to drink.

Not that it made much difference. First my mouth and later my whole body absorbed as much water as I could put into it and I was still thirsty. I was thirsty when we rode into the mission compound and hung about while Raimundo exchanged the elaborate nothings of greeting with other peons. Eventually we

Shavante: Men dancing
while the children watch

Shavante: Women sitting
round an earth oven (*above*)

Pia and Biorn sharing huts
with the Shavante (*right*)

Shavante: Dancing with
masks

were directed to the refectory, and on our way we met a young priest with a long, fair beard and the eyes of a conquistador. He was striding along in his white cassock as if he had an important appointment to keep. Later I tried hard to remember what was so striking about this first chance encounter with him and eventually I realized that it was that brisk walk. In Mato Grosso men tend to walk slowly and economically, sparing themselves by drawing on their endless reserves of time.

He received us courteously but coolly, the whole set of his body proclaiming his willingness to offer us the customary hospitality and his disinclination to give us charity. I was suddenly very conscious of my tatterdemalion appearance, my floppy straw hat and parched, stubbly face, my wet trouser legs and sodden tennis shoes. Hastily I explained that I was the Englishman who had been working among the Shavante, thereby establishing two unimpeachable claims to eccentricity. Our host's manner changed at once. He was glad to see us, he told me, and hoped that I and the mission could derive mutual benefit from my stay in Sta Terezinha. They had only been working with the Shavante a very short time and would appreciate information about them from someone who was qualified to give it. He strode ahead of us as he talked, leading the way to the mission buildings. I very much doubted whether I was yet qualified to give useful information about the Shavante. I had just about reached the stage when I could fully appreciate the limitations of my own knowledge. On the other hand, I did not feel like voicing my reservations there and then. I just wanted a drink of water.

As soon as I got to the earthenware water jar I settled down to a drinking debauch. It is hard to convey to those who have not experienced it the sensation of drinking till one is utterly water-logged and yet still feeling thirsty. I drank until my stomach could hold no more and as a result I no longer felt hungry, not until I joined the missionaries for supper that night. We were served a thick, Italian soup made with the vegetables from their own gardens, and I had not seen fresh vegetables since leaving São Paulo.

Sta Terezinha had many of the attractions of an oasis. There were two priests and three lay brothers at the mission, and if they had not yet succeeded in making the wilderness bloom like a rose, they nevertheless raised livestock, constructed buildings, set up workshops and (above all) grew vegetables. In fact, they were embarrassingly active. They had built a convent for some nuns

who had arrived to school the Shavante and they were busy digging watercourses to provide power for further projects. Day in
and day out they toiled in the hot sun, and I soon lost the feeling
that it was incongruous to see a man in a white cassock swinging
a pick in a near-by ditch. What I did feel was incongruous was
that they would not permit me to help. I thought it only reasonable that I should do something in return for their hospitality,
but they were adamant.

'Scientists,' they insisted, 'do not do this sort of work. They
must get on with Science.'

It would be nice if anthropologists in all parts of the world
were similarly relieved of the chores of life, but here in the middle
of Mato Grosso it made me feel ill at ease. When the table was
cleared after our morning meal and the missionaries all sallied
forth to continue their battle with the environment, I was left in
state with my papers and such books as I had. Under the circumstances, it was not always easy to decide how best to advance
the cause of Science, but I felt morally obligated to do so.

First, I sought out the local Shavante and discovered to my
surprise that their village was only a couple of hundred yards
from the mission. There were even seven huts in it, one only half
finished, and I began to suspect at once that I would have to deal
with 'mission Indians' with all the unpleasant connotations of
dependence, uncertainty and lack of self-respect which the term
so often implies. The first Shavante I met confirmed this impression. They were youths who swaggered about in their city-
made snakeskin belts and told me at once that they had nothing
but contempt for the naked savages where I came from. They
would probably have given up speaking Shavante if they had had
sufficient command of Portuguese, but here again I was disappointed. They had managed to disassociate themselves from
their people without acquiring enough knowledge of anything
else to make them valuable as informants. They trotted out their
few sentences of pidgin Portuguese like European teen-agers
airing the latest Americanisms.

Later on I discovered that they were not as *civilisado* themselves as they would have liked me to think. After they had shown
me their bright shirts and the shorts which they had earned from
the mission, they caught sight of the bead necklet I was wearing.

'Who is your uncle?' they demanded at once, for Shavante boys
are traditionally fitted out with necklaces by their mothers'
brothers. I told them that I had been given the necklet by my

brother-in-law Suwapte and that it was he who saw to it always that I went about properly dressed. But the boys at Sta Terezinha hooted with laughter at the idea that a man might receive such a necklet from his brother-in-law.

'Give it to me,' one of them demanded immediately.

'I cannot. My brother-in-law will be angry when I go back. He will say—Where is your necklace?—and then what will I answer?'

They roared with laughter again.

'He will not mind. Brothers-in-law are always angry. Give it to us.'

I refused.

'But there are no beads here,' they remarked indignantly, as if that clinched the matter.

I had watched these beads being made out in the Roncador and I knew what a painstakingly laborious business it was to collect the plants, shake out the seeds, burn them (but not too much) so that they hardened and then pierce each single one with a bone needle. It required days of work for a single necklace.

'There are no beads here because you are too lazy to get them,' I countered.

Their faces clouded over with sulky outrage.

'No. There are no beads. There are no beads here. We work hard but we have no beads.'

'Then why don't you go on trek and get some?'

'There are no beads to be had where we go,' their voices rose to a shrill note of protest.

'Of course there are,' I argued, deliberately irritating them to see what they would tell me. 'We got these beads in the Roncador. Why cannot you get some the other side of the river too?'

'Because there are NO BEADS where we trek.' They spelled it out for me as one might to a particularly exasperating imbecile.

'Then why don't you trek over to the Roncador and get some?'

It was the question I had been leading up to, and they were so heated that the answer came before they had thought about it.

'*Pipa-di* (frightened).'

'Frightened of what? They won't kill you.'

'Of course they'll kill us. They kill lots of people. They would kill us too if we went over there. *Pipa-di*.'

'Why should they want to kill you?'

'Apewen is a bad man and his people are bad people. They kill many. That is why he is a strong chief.'

I tried to get them to elaborate this point a little, but they felt already that they had told me too much.

'He is a strong chief,' they kept reiterating. 'He kills many.'

I thought of S'rimri and the shorn heads of his wife and her mother. Had he been sacrificed so that Apewen could acquire strength? I found it hard to believe. S'rimri was so slow that I could not imagine him posing a threat to Apewen. But the fact remained that he had been killed and I still did not know the reason why.

Then one day I played them a tape which I had recorded at the other community. It was half an hour of the discussion which took place one evening in the mature men's council and try as I would I had not yet been able to decipher all of it. I knew it dealt with the plans of Apewen's people to wander southwards on a hunting expedition, but I could not hear the details clearly enough to understand them. The effect it produced was out of all proportion to what I expected. The boys were first astounded and then fascinated with a horrid fascination that even communicated itself to me.

'What is it? What are they saying?' I kept asking, but they shook their heads and motioned me to keep quiet.

When the tape was played out they unanimously demanded that it be run through again. Here I insisted that they first tell me what was so interesting about it. Only then did I get the story.

It was a discussion of an excursion to the south all right but it had also included at one point a suggestion that this might be combined with a raid on Sta Terezinha.

'What for?' I asked.

'*Meh*, they are angry with us, always angry with us.'

'Then why didn't they attack?'

'They were afraid.'

'*They* were afraid? But you have been telling me that you are afraid of them!'

The boys looked uncomfortable.

'They were afraid of sorcery,' they explained at last, lowering their voices, 'but there is no sorcery here. Here we live in peace. There they are always killing.'

'But they are *not* always killing,' I was getting heated myself now. 'Tell me who they have killed, if they are always killing so much.'

'Listen to the talking-thing,' one of them shouted, 'just listen. They always want to fight. They are always angry.'

'They killed Reamzu, didn't they?' chimed in another.

'How about Serewede?' added a third.

'Listen to the talking-thing. Turn on the talking-thing.'

I was happy to comply with their request. I had to have time to think out the implications of all this and to jot down the names of the men who had apparently been rubbed out in my community. Sorcery and killing, killing and sorcery, these seemed to be the keynotes of the discussion. Gradually the whole story came out.

Reamzu and Serewede had been killed about five years ago when Apewen's group was trekking through the Roncador. They were not the only ones either. It was said that as many as eight people had been slaughtered in a single night, though people's recollections of the carnage were confused and contradictory. It all happened because Apewen's brother died. Sorcery was suspected and Apewen's people therefore singled out a number of suspects and murdered them in their huts. Many others, kinsmen of the murdered men, fled precipitately lest the same fate overtake them, so that overnight a sizeable proportion of Apewen's band scattered to other Shavante communities.

'Then how is it that he is considered such a strong chief now,' I asked, 'if so many of his people have gone to the other villages?'

'That was some time ago, and besides Shavante come and go. There is always quarrelling here and intrigue there. Communities grow big and then there is a fight and they split up again.'

I understood now why Shavante villages were usually on bad terms with each other. There were always refugees at each place, breathing venom at the band from which they had fled.

'How about S'rimri?' I asked as a parting shot. 'Why did Apewen's people kill him?'

'Because one of them seduced his wife,' came the paradoxical reply.

S'rimri had been killed by the people who had seduced his wife! I knew there was no enmity like the enmity a man bears towards someone whom he has wronged, but this seemed to be carrying it a little far. Besides, the killing of S'rimri was closer to an execution than a murder. Moreover, the Shavante do not regard being cuckolded as a crime; on the contrary, they are extremely touchy about the husband's rights in such matters. I could have understood it if the seducer had been executed, but the wronged husband . . . and then I remembered the close relationship between the seducer and the chief. I shuddered. This new slant on Shavante society did not make it appear very cozy.

The more I probed, as the days went by, the less cozy it appeared. I discovered that recently about fifty Shavante had left the Sta Terezinha community and gone to join one of the other villages upstream. They too had been on the losing end of a dispute with their chief and had gone so far as to accuse him of killing his own wives (their kinswomen) by sorcery. Now too I realized why Suwapte had appointed himself my patron and why Urbepte had formally become my friend. They were of different clans; Apewen's clan was yet a third one. If Apewen adopted me, then Suwapte and Urbepte were probably anxious to see that I was not exploited to the political or economic advantage of a single faction in the community.

By now I was longing to get to the unknown Shavante of Marawasede and find out why there were always refugees fleeing from that band over to Apewen's people. I felt sure that if Pia and Biorn came too and we were able to drop in on the community by air there would be no danger involved.

So the following Thursday I packed my rucksack and made ready to leave on the supply plane from Campo Grande. It had called regularly during my stay at the mission and I had every reason to suppose that I could be in Campo Grande by nightfall. But that Thursday was the first one it missed. We did not think too much of it. A plane may be days late in the interior if it develops engine trouble. When Friday and Saturday passed and it still had not come even I was forced to abandon hope for the week. I unpacked my things again and did my best for Science. But the Shavante were beginning to bore me, and every time I tried to analyse the information in my notes I was frustrated by some point which could only be cleared up by working in another village. I recognized that my inability to concentrate stemmed from my impatience to get away from the mission, so I decided to tackle the problem frontally by stopping work for a few days. 'When you have to wait, relax' could be the motto of the interior, and so I did my best to relax. When the plane missed again the following Thursday, however, I began to find relaxing hard work.

I was pleasantly but effectively trapped. The missionaries were kindness itself and sought to cheer me up by telling me that in all probability the air base had discontinued the landings at Sta

Terezinha because of bad visibility at this time of year. Everybody was burning off their gardens for planting at the first rains and smoke hung in a thick haze over the entire state.

'As soon as it clears, they'll come again, don't worry,' I was told. 'Until then please regard this as your house.'

'But how long will it take to clear?' I asked.

'Who knows? Sometimes a week or two, sometimes a month or more. . . .'

There was no other way out. The mission had a motor-boat but they had run out of oil and could not operate it until the plane brought them some more. Nor could they spare any of their workmen to go off on a four- or five-day journey to guide me overland to Shavantina. I had to possess myself with as much patience as I could muster.

I kept my rucksack perpetually packed so that I only needed to sweep my books together and roll up my hammock to be ready to leave. This preparedness paid off, for one day a plane did appear unexpectedly and, before anybody realized it, was circling to land. I ran faster than I had run for a long time to make sure that the pilot did not take off again in a hurry and discovered that it was the Bishop himself coming to visit his outlying missions. I am afraid I did not greet him as formally as I perhaps should have done, for I was anxious to have the pilot's word for it that he would take me to wherever he was going and furthermore that he would wait for me until I had time to collect my effects. A few days later I was back in São Paulo.

The DC-6 seemed a leviathan after the little planes of the interior. When it finally rolled to a stop in front of the glass vista windows of Congonhas Airport I felt both elated and ill at ease. I took my bags past the hostile city stares of all those people that just hang about in terminals and tried to find a telephone. I had no idea of where to go. I did not know where Pia and Biorn were and, as luck would have it, this was a Saturday. The consulate was closed and our friends were probably away for the weekend. For all I knew Pia might be away for the weekend too. I did not have the right change for the telephone and every time I tried to get some I received the same looks—pitying looks directed towards the peasant who does not know his way around the big city. I phoned and phoned without success. Then I thought of the Danish consul and his English wife who had been so kind to us. I called their house and she told me that she knew where Pia was staying but that her husband had got the actual address written down and he was not in. Why didn't I come over and have a meal with them?

It was an offer I was only too glad to accept. My hostess may have regretted it for she later insisted on driving me round to Pia and Biorn. We were overtaken on the way by one of those convulsive rainstorms which battered at the car and made the road into an agitated lake. Since the road also happened to be under repair, we found ourselves suddenly with the car heeling over at a drunken angle and stuck fast.

'All the way from Mato Grosso—and now this,' I could not help thinking as I clambered out and found myself up to my calves in water. I could not move the car. My hostess climbed out gamely and promptly lost a shoe. The rain was still pelting down when a little man appeared, soaked to the skin, and offered to help. Between the three of us we refloated the vehicle, but when I turned round to offer our benefactor something for his trouble, he had disappeared. I felt certain that he too was from the interior.

12

We and they

There was to be no helicopter. They were all but unobtainable in Brazil except from the Search and Rescue Service, who were understandably reluctant to send one of their machines off into the wilds to look for Indians.

Pia and I decided that we would take Biorn with us and return to the Shavante via the air base at Campo Grande. It was a difficult decision to take. Our friends, to whom we were by now so deeply indebted, branded the proposal as foolhardy. Our doctor was equally emphatic. He went so far as to say that we would be responsible for the child's life if he sickened again. But we were the only people who really knew what we were taking him back to and we were convinced, in spite of everything, that the risk was not so great and that it was justifiable. We were returning to the final weeks of the dry season when the climate is at its best. With luck it might prove a good convalescence for the child. If we were unlucky . . . we decided we would not be unlucky.

A few days later we climbed into yet another Beechcraft which flew us to Pimental Barbosa. Once again we stood on the familiar strip of rank grass, our bundles about our feet, and peered into the trees for any signs of life.

'Where are the Indians?' asked Pia, surprised.

'I don't know. I wonder why nobody comes from the post either.'

There was a tiny thatched shelter at one end of the strip where people could leave their belongings or, if necessary, take refuge from the harsh embrace of the sun. I took all our things over and put them there.

'Let's go over to the village and see what's happening,' I said.

Here I could carry the baby, for Indians did not feel so strongly about this being inappropriate for a man, and the personnel up at the post had long ago come to the conclusion that we were unaccountable eccentrics from whom anything could be expected. I lifted him on to my hip and Pia took the rucksack. As we moved off a couple of Shavante came into the clearing. They were Waarodi and a boy who had recently arrived from Marawasede.

Their welcome was not exactly effusive. They asked us what we had brought from the city and I told them that there were many things in our boxes, indicating the shelter.

'Let's see them,' ordered Waarodi.

'*Hadu*—wait.' We were back in the old conversational groove again. 'You will see them when they have been brought to the village.'

Waarodi was as usual disinclined to bring anything anywhere and his companion was too busy playing with Biorn.

'Where are the others?' I asked.

It was their turn to say '*Hadu*'. We were told that they had not yet got back from their trek. So at least we had not arrived late in spite of all our setbacks.

'Where is their smoke?'

'Over there.'

'Much smoke?'

'Mu-u-u-ch.'

'Far away?'

'Ve-e-e-ry close.'

'Sounds as if they will be here in a day or two,' I said to Pia.

'We seem to have timed our arrival nicely.'

Waarodi and the boy accompanied us back to the village, carrying the two lightest things they could find in our baggage. The great half circle of huts was as forlorn as a row of empty hives when the bees are swarming. There was not a single Shavante in sight. Waarodi deposited what he was carrying for us and stalked off to his own hut. A moment later I knew why. Our dance masks were gone. It had taken me weeks of bargaining and cajoling to collect nine of them for three separate museums. When the village went on trek the elders came in and bound them carefully around a single pole so that they might be stored without deteriorating. At that time I had wondered about the advisability of leaving them in my hut while I was away, but Apewen assured me that they would be perfectly safe.

'Waarodi is staying to look after the village,' the old man told me. 'He will watch over them. Besides, nobody would steal such things.'

I was fuming as I went across to Waarodi's hut.

'Where are my dance masks?' I demanded.

Waarodi feigned first ignorance and then surprise. Weren't they in my hut? Then he would come to look.

The two of us went back across the deserted village. He came into our hut and stood there with his chest chucked out like a pouter pigeon. This was his normal stance. He looked all round the hut and even went through the performance of ferreting for something, which irritated me exceedingly for there was nowhere where a seven-foot pole swathed in dance masks could possibly have been overlooked in our floor space.

'Who took them?' I demanded.

He thought for a while.

'The spirits must have taken them,' he replied at last.

I was dumbfounded. I had been prepared to deal and indignantly with almost any other reply. But the spirits . . . Furthermore, he had named a special category of spirits, the *wazepari'wa*, the very ones which my age-set had impersonated at the end of the initiation ceremonies, who were associated with the marriage tie and thus with these dance masks. I could do nothing but stammer my disbelief.

'Spirits nothing. People have stolen these masks and I want to know who they are.'

'Not people, spirits,' replied Waarodi imperturbably. 'There have been no people here.'

I felt as frustrated as a lawyer faced by a dishonest but unshakable witness. That was his story and he would not budge from it.

'Of course he took them himself,' said Pia as soon as he left. 'He probably sold them to the crew of a passing plane if there was one. Somebody is quite likely to have landed here since you went away.'

'But how on earth can I prove it?' I moaned. 'Just wait till the others get back. I'll tell Apewen and see if I cannot force a showdown.' But as I said it, I knew that I spoke more to relieve my own feelings than to convince anybody. I was not likely to fare well if I tried to accuse Waarodi before his people.

Exactly how badly I might fare was underlined by the events of the following afternoon. Pia and I had slung our hammocks in the verandah of the post and even Biorn had finally fallen asleep, overcome by the heat. It was one of those days when one could speak of 'biting heat', so sharp was the impact of the clear sunlight on people and things alike. As we rocked and talked a shrill halloo came echoing across the river.

243

'They're back!' I said, jumping out of my hammock, and it was not till I was already out in the smarting sunshine that I remembered that 'our' band were not trekking the other side of the river. Still, I thought, some of them might have crossed over for some reason or other and they would be sure to want a canoe to go over and fetch them. I could only make out a single figure on the other side. I could not recognize him at that distance, nor as we shouted to each other, could I recognize his voice. He was so obviously surprised to be answered in Shavante by a white man that I knew at once he was not from our community. He sounded very flustered. When Shavante halloo to each other in this way, their voices run up the scale with each sentence till they reach a high falsetto which carries over great distances. Then, at the end of each phrase, they drop back to a normal or perhaps unnaturally deep note which functions as a period in their conversation. This man's voice cracked on the high notes and he could not manage the stops at all. He was obviously out of breath and very frightened.

As we shouted to and fro, Pia and Biorn came out to see what was happening as did a lone Shavante who was hanging around the post.

'Take the canoe over and get him,' I told the young man.

The single paddler made poor headway against the strong current so that it seemed to take him an interminable time to reach the opposite bank. At last he got there and the newcomer stepped into the boat. But they did not start back at once. The youth who had gone to fetch him recognized the visitor as a member of his own lineage, so he wept for some minutes over him as custom demanded before starting back for the near shore.

By this time other Shavante were hastening up from the village, as though informed by their sixth sense that something was afoot. Waarodi was there, of course, and Sibupa whom we had not seen for months. He greeted us effusively.

'Where have you been?' I enquired. 'I did not know you were here.'

'Out on trek,' he replied.

'I didn't know. Is everybody back from the trek?'

'Only a few. The rest come tomorrow.'

The stranger was getting out of the canoe now, and Sibupa assumed the expressionless countenance appropriate to the occasion. I noticed that the new arrival had clots of blood on his shoulder and decided at once that I would have to go and hear what he had to say. Waarodi and Sibupa escorted him in single file back to the village without a word being spoken. I brought up the rear. One or two people had come back since the previous day but they afforded me no sign of recognition and I ignored them. Solemnly we marched over to Waarodi's hut, as if we were escorting a prisoner.

The visitor lay down on a sleeping mat and the others gathered about him to weep. Men and women, they bowed their heads and wailed till the harsh yet plaintive corncrake sound of their weeping filled the whole hut. I sat like a ghost with no place in their ceremony, watching fascinated while the tears dripped on to the weepers' chests and strings of mucus hung from their nostrils. One by one they stopped their wailing and cleaned their faces with blades of grass. Then the men lay down beside their visitor and for a while nobody spoke.

At last Waarodi began to talk. He spoke in the formal manner but so softly that he might have been an old man pronouncing his dying oration. As far as my straining ears could tell he was asking the visitor his reasons for coming and inviting him to give his news. Then the newcomer explained that the people of Marawasede were incensed with him. He had run away in order to avoid being killed but they had pursued him along the trails. For days they had tracked him and fired arrows at him as he ran. They had wounded him in the shoulder and in the leg but he eluded them and managed to reach the Old River (Rio das Mortes). He had not slept once on the way and had eaten only what he could gather in his flight.

Next day the villagers returned from their trek, and the refugee had to repeat his story in the men's council. As he was speaking, Suwapte got up to be his complementary speaker. I was momentarily surprised that Apewen did not undertake the opposing speech on such an important occasion, until I remembered that Suwapte and all the fellow members of his lineage had themselves come from Marawasede some time back. He was clearly the expert on the matter in hand. It was simple enough on the face of it—all their brothers were being killed off by the Marawasede Shavante. He was one of the few that had escaped. There were murmurs of assent from the circle of men, and three or four of them started making speeches on their own account from where they sat. I was doing calculations in my head . . . yes, at least six men had come to us from the Marawasede group since the start of my field work.

I was beginning to see the pattern of it. The chief of Marawa-sede was dead, and since his death there had been killings in that band. All the men who came fleeing to us were members of Apewen's lineage, and they brought news that their brothers were being massacred where they came from. It looked as though the lineages were battling as a result of the old chief's death—the refugees were all held either to have been implicated in killing their chief by sorcery or were kinsmen of others who were so implicated. In fact, there was a purge going on in Marawasede similar to the one which Apewen had successfully carried out in his own community some years ago.

Once I knew so many of the details of these cases it was comparatively easy for me to probe for the rest. It was a strange sensation to lie on their sleeping mats through the long afternoons while Biorn and the children of the household (and their pets too if they had any) wrestled and danced in the corners of the hut and listen to these tales of murder and intrigue, haltingly and surreptitiously told. I was left under no illusions as to the fact that this was dangerous information which was being imparted. My informants wriggled with embarrassment and lowered their voices. Sometimes they refused to speak in the presence of their wives, sometimes even in the presence of their children.

Yes, it was true that Apewen's people had killed many while they were out in the Roncador back in the time when the last age-set were in the bachelors' hut (five years ago). Yes, it was when Apewen's brother died. Apewen had ordered that the dead man's son-in-law be killed as a sorcerer and also the fathers-in-law of two of his own sons. Three men related by marriage to his own kin and the kin of these men to a total of eight. I began to understand why it was that the Shavante believed that malevolent spirits presided over the marriage bond. I also understood now why it was Apewen's clan which had to act against these men. They were engaged in wiping out another lineage and so they obviously could not get the whole community to agree to take joint action against them.

The case of S'rimri was different though. He had been killed by the community. I got the full story from the mother of Surupredu, the man who had led the executioners. Tearfully she took me aside and asked me if it was true as the Indian agent said that the 'tying-up people' (soldiers) would come and take away any Shavante who killed another one. I hardly knew what to say. I thought that it was highly unlikely, but it was one of the few threats which the agent had in his armory of defences against the Shavante. I compromised by saying that I knew they did that in Shavantina and they would probably come here too if they heard about any killings. She begged me not to tell them, for her boy was a good boy and S'rimri had clearly bewitched him.

'Why did he bewitch him?' I asked with renewed interest.

'He was angry. He said my son went with his wife and that he was not going to stand for it.'

'Then what happened? How did you know he was bewitched?'

'He got a sore on his leg. A big sore which would not go away. You saw how he walked when he came back from the Roncador. Who else would have done that to him but S'rimri? Besides it was not only my son who was affected. His brother Mantse also fell ill, so S'rimri was bewitching the whole lineage.'

I reflected that it was exceedingly stupid of S'rimri to air his grudge against the dominant lineage in this fashion and doubly stupid not to watch out for the possible consequences once he had done so. But then I already knew how stupid he was from personal experience. His uncle had been much more perceptive. The day the warrior age-set was detailed off to perform the execution, he fled up river and was not seen again.

So it was with trepidation that I aired my particular grudge against the dominant lineage, or rather against its second most influential member, Waarodi. I did not mention his name, but I told Apewen in private about the theft of my dance masks and brought the matter up again in the men's council. There ensued a heated discussion about the various possibilities in the case and the final verdict was that they had indeed been spirited away. From this there was no appeal. I think that the majority of the men believed it to be true, since according to Shavante these masks were the property of those particular spirits who were thus simply held to have claimed their own. I was equally certain that there were some who knew very well that this was not true . . . but I did not feel it would be wise, under the circumstances, to insist.

I had to set about collecting more masks, which was a difficult and wearisome business for the supply was not unlimited and they were valuable property, not lightly parted with at the best of times and especially not now that the spirits had intervened so dramatically with my first consignment. We had brought more presents with us from São Paulo and these helped, but Pia and Biorn helped more than anything else.

The Indians had been delighted to see them return. For the first time in the whole of our stay among the Shavante I began to feel that our presence was not irksome to our hosts. Our hut had always been full of children at all hours of the day and night, but now adults wandered over to see us when they had nothing better to do. Sometimes they would stay there all day. At others, they would stay all night as well, till our tiny hut began to feel more like a caravanserai than a home.

Biorn enjoyed this and the opportunities it gave him for showing off. We were usually embarrassed by his rumbustiousness, compared with quiet Shavante children. He invariably seemed to be shouting more than all of his playmates put together and they put up with a great deal of shoving around from him. It was not that they were more severely disciplined. On the contrary, they were allowed to do more or less what they wanted. If they were fractious, they were just ignored. We saw little boys indulge in tantrums lasting anything from ten minutes to half an hour. One lay biting and pummelling the ground with rage and frustration, another stood up with his toy bow and gave a hilarious imitation of a senior man haranguing the village, only he was screaming his outrage at his mother and sisters. Nobody took any notice and eventually they calmed down and went back to playing. But such outbursts were rare. Usually the children played with less shouting and screaming than Biorn did. Sometimes they were so fascinated by his antics in fact that they gathered to watch him as if he were some strange animal, which in a sense he was.

The result was not only that people were always in and out of our hut but also that now I was made welcome in huts where previously I had rarely ventured. There were days when I followed Biorn from hut to hut, while Pia was away on collecting expeditions with the women or while she lounged with them hour after hour, in the limpid waters of the creek.

At this stage of the dry season the women and children seemed to spend just about the whole day in the water. It was the pleasantest place to be and also the most sociable, but it made it hard for me whenever I wanted to swim. In a community of naked or near naked people I was the only person who could shock them with my immodesty, and this was because I did not wear the tiny penis sheath which all adult Shavante men affect. I could have done, I suppose, but apart from it being very uncomfortable I felt that it would be a usurpation of Shavante-ness which it was

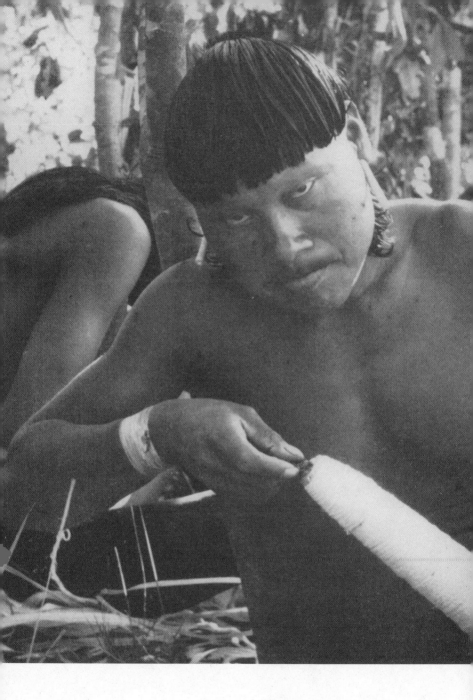

Shavante: Disgruntled

Author with his son and refugee from Marawasede (*above*)

Pia and Biorn packing to leave the Shavante (*right*)

Shavante: Apewen, the
strong chief

not up to me to assume. If they chose to suggest that I wore one that would be another matter. Until then I had to dive for the bushes every time women came down to swim, whereas nobody minded if Pia threw off her clothes and bathed with the men and women who were generally splashing around when she got there.

Apart from this handicap, which gave me an unexpected insight into some of the disabilities which the requirements of modesty impose upon women in our own culture, it was an idyllic existence. We almost forgot the grim side of Shavante life —almost. I was reminded of it again, and forcefully, when we heard it rumoured that Suwapte was a sorcerer. It was no less startling when we heard it because we had predicted that such a thing would happen sooner or later. Suwapte was the chief's son-in-law and, therefore, in a delicate position. He had also earned the enmity of the chief's eldest son by showing himself a frequent and impressive speaker in the men's council. Finally, he was the senior member of the lineage which had arrived from Marawa-sede, people who were not even of the same clan as the chief's. He must have seemed a dangerous rival to Apewen's sons.

These were tense days. We knew that Shavante often talked round and round a dispute for weeks. In fact, they very often solved them by 'talking them out' in the men's council so that tempers cooled and nobody could be bothered any more. But we also knew that they were exceedingly volatile people who might unexpectedly leap for their clubs. I was not anxious for that to happen while we were there. We had heard enough descriptions of Apewen's purge to want to be out of the way whenever there was a repetition of it. I thought of the young man who was nick-named 'Pipa-di' (frightened) because he had run terrified from the village when the killings took place and was later found dazed and wandering in the woods. He must have been about thirteen years old at the time.

So we awaited developments, but nothing seemed to happen. The nights were bitterly cold now and there were always jaguar snuffling and roaring round the village, possibly because they were lured by the cattle at the near-by post. The Shavante seemed more concerned with them than with their own disputes. Hunting parties occasionally went out after them, but they never killed one, though the skin of a huge jaguar outside Apewen's hut testified to the fact that they were not incapable of doing so.

One night Pia awoke to hear an *onça* just outside our hut, or so it seemed. Never had the cozy thatching seemed frailer to us.

The beast must have been pretty close, for a moment later one of Apewen's sons from the next hut crept in to us.

'Apewen,' he said to me, 'where is your gun?'

'There,' I pointed to my heavier rifle.

'No, no,' he insisted, 'the one with big bullets.' I knew he meant the revolver. The Shavante had been deeply impressed by the girth of the ·38 slugs. I showed it to him.

'Let's go,' he exclaimed.

'What for?'

'To kill the jaguar.'

I was unenthusiastic. Surupredu came in now and added his entreaties. They did not even know where the animal was but they assured me that it was 'very close'. I knew the elasticity of 'very close' among the Shavante and had no desire to wander blind into the darkened forest with an inaccurate blunderbuss of a weapon to be stalked by some jaguar who seemed unhealthily familiar with men. The Shavante were angry.

'Give us the gun, then,' they said.

I gave it to them, gingerly, and taking care to keep it pointed well away from us all. The two of them disappeared, and we tried to get back to sleep again. We were still listening for the jaguar and wondering whether the young men had really gone out after it by night when we heard the unmistakable crash of my revolver. Once, twice, three times . . . I was out of my hammock now and going for my rifle. By the time Pia and I emerged into the darkened village we had heard all six chambers discharged. I imagined Surupredu frantically emptying the gun into a charging jaguar, but there were no shrieks or growls or other signs of life. Only a mutter of talk from Apewen's hut and some Indians slipping like shadows into the woods beyond. I followed them and met Surupredu.

'Why did you shoot?' I asked him.

'To frighten the jaguar.'

'Why so many shots?'

'Makes a big noise. Frightens the jaguar. . . .' he replied, with his eyes shining. 'I want more balls for the gun.'

Surupredu apparently thought it was some sort of firework. I was indignant.

'I have not got so many balls that you can just shoot them in the air,' I remonstrated. He was unrepentant.

'Makes a big noise,' he kept saying happily.

Biorn was not only awake but tangled up in his mosquito

netting and crying bitterly in the darkness when we got back to the hut. We both of us cursed Surupredu and his big noise.

Next morning we were out looking for jaguar again. I came back at midday, and in the afternoon I lay in my hammock trying to write notes through the interruptions of the row of little boys who were taking their siesta right under me. Pia and Biorn were dozing restlessly when Suwapte came in, carrying an axe. There were only one or two of these in the whole village, and I remember wondering what Suwapte was going to do with this one when he came over to my hammock and swung it up above his head as if to bring it smashing down on my face. I saw his features contorted above me as one might on a less tasteful advertisement for a violent film and I thought at that moment—'But of course, he's my brother-in-law!' In my hammock I was completely powerless to dodge. I don't think I so much as moved. He gave me a broad grin then and pretended to hack my head off with a terrific flourish. It was his idea of a practical joke. Luckily I had not been in a position to give him the satisfaction of seeing me take it seriously. He sat down more or less on top of me, for my hammock was not made for two, and asked me for a present. As we talked I remembered that I had seen other Shavante play similar practical jokes before. I had seen them leap up and take a bow and arrow and aim at the heart of some three-year-old who would promptly burst into terrified sobs. Theirs was a macabre sense of humour but I should have got used to it by now.

Suwapte told me that he was going out to get saplings for a new hut.

'But what's wrong with the old one?' I asked.

'It's a bad hut,' he said. 'Soon there will be rain and then we must have a good house.'

I went outside and examined Apewen's house critically. It looked all right to me, but it was clearly the season for house repairs. I could see another household stripping their hut on the far side of the village and carefully piling the thatch so that it could be used on the new house. For the next few days Suwapte was in and out of the woods gathering trees. Apewen helped him occasionally, but his long row of sons lay like so many crusaders on their sleeping mats and did not lift a finger. This was a job for in-laws only. Seeing which, Pia mentioned that they would probably put up with Suwapte's sorcery at least until the house was built.

That night it rained. It was only a light shower, but indicative of the change of season. A few days later it rained in earnest and we found out why people had been rethatching their huts so assiduously. We watched the storm clouds gather in the hazy sky and wandered cheerfully back to our hut, if anything rather looking forward to the onset of the rains. But we were utterly unprepared for what was to follow. We noticed the urgency with which the normally haphazard Shavante were scuttling to their houses and battening down the thatch, but before we had time to consider the implications of their actions a hot wind whipped across the centre of the village, gathering every speck of dust and particle of loose straw and flung the whole lot in our faces. I picked up Biorn who was crying and rubbing his smarting eyes and we stumbled blindly into the hut. There was havoc inside. The wind had blown over the Primus and scattered our pans. It had caught our hammocks and tossed them in the air until they were trapped by protrusions in the thatch or from our home-made wooden shelves. My notebooks were eddying about my feet, my writing materials already half buried in the silt that was being blown in at us as if it came out of the wrong end of a vacuum cleaner.

I pulled down Biorn's hammock and popped him into it, just to weight it down. Pia started rescuing the things on the ground and I struggled to block the entrance against the wind. It was not an easy thing to do. I had to gather loose bundles of thatch, place them across the entry opening and weight them down with two logs. The logs, of course, leaned against the outside. I was on the inside. All this in high wind which tended to take the whole thing out of my hands and slap me in the face with it.

The first slanting spikes of rain came before the wind dropped. I was still trying hard to adjust our ineffectual 'door' when they came driving at me like needles. The whole hut was fluttering by this time with each gust of wind. I had only just got the doorway blocked when an especially strong blast lifted the roof off. A huge patch of our overhead thatch vanished into the jungle and simultaneously the rain poured through into our meagre floor space. Once again we had to rescue our belongings. Biorn was whisked out of his hammock and bedded down on the palm leaves in a dry corner. Notebooks and cameras were hastily retrieved and put under plastic. Then we stood and looked at the breach in our defences. The rain was coming down really hard now and the wind was dropping slightly, though it still ruffled the hut now and again. I felt like Lear roaming on the heath.

'If the whole hut blows away, what will we do then?' said Pia, voicing my unspoken thought.

'You take the baby and I'll take the notebooks,' I replied.

'And where will we go with them?'

'I don't know. To one of the other huts I suppose.'

I wished the Shavante had thatched ours with a little more care, but in my heart of hearts I could not blame them. They could not know how long we planned to stay with them, and in any case they had no reason to give us the full value of their craftsmanship. Our hut was, frankly, jerry-built, and the prospect of spending quite a lot of the rainy season in it was uninviting.

I took away my door-stop and edged outside into the storm to see how other houses were faring. It was some small consolation to notice that not all of them had come through unscathed. There were people carrying out emergency repairs on the thatch of quite a few of them. Apewen was standing outside his and directing the efforts of his daughter (Suwapte's wife) as she repaired a hole in the fabric. I pointed out that ours too had been damaged and he laughed with the excitement of it all, the rain dripping all the while off his aquiline nose.

'Sister,' I called up to the girl on the roof, 'come and make my house too.'

'After the rain,' she replied, shortly.

At that moment I saw a young man trotting towards us through the rain. Duinwe was something of a simpleton. He was neither so well built nor so fleet as the others of the warrior age-set and he tended to act the clown whenever he was called upon to join them in their ceremonial activities. So he was ignored, if not despised, but at the same time he was nobody's enemy. I clutched at this straw of hope.

'Duinwe,' I called, 'come and make our house.'

He stopped and looked up at the hole in the roof.

'Make our house,' I went on quickly, 'and then we can eat.'

That did it. He went off and gathered some thatch and clambered up to mend our hole. It was an amateurish job, for men do not normally thatch among the Shavante, but it kept most of the rain out. The hut still leaked with a steady drip, drip, drip in several places but at least we could retreat to the dry corners without being cascaded upon. Duinwe came in, and we lent him a towel to get dry, which amused him for Shavante always let water dry on them. He asked if he could keep it and we gave it to him for his kindness. Then Pia brewed some cocoa to cheer us all up. Duinwe was overwhelmed. He had never tasted cocoa before and he definitely approved. From that day on he hung about our hut, hinting that he would like some of that sweet drink we had given him and he became our unofficial odd-job man in exchange for a regular supply of cocoa. Being a simple soul he was unremittingly patient with children and he would sit playing with Biorn by the hour when we were obviously disinclined to brew cocoa for him there and then.

He appeared to have one immediate ambition—to acquire a torch. We promised him one of our torches when we left. I did not dare to give it to him before then for fear of arousing the jealousy of the others. They would be offended if Duinwe, a person of no consequence, were given a valuable gift like a torch when they themselves were not offered such a thing. But it was hard for us to keep him waiting. We gave him other presents of course, but it was the torch he really wanted most of all, and with a pathetic insistence. One day I asked him, simply to play for time, why he wanted a torch so badly when I offered him so many better things. It was intended as a rhetorical question, for I already knew why Shavante valued torches—they wanted them for hunting by night. His answer came as a shock. He wanted a torch, he told me, in order to frighten away the *wazepari'wa* spirits when they prowled around at night. I had only once before heard a man give this as a reason for desiring a torch, and I had dismissed it from my mind as a colourful invention designed to get me to part with valuable property. Now Duinwe gave me the same answer—and the two people who gave this reason slept in the same hut.

It was a little girl who solved the mystery of the spirits for me. She and her two sisters were nearly suffocating me with their attentions one day, for they never tired of pulling the hair on my legs, feeling the skin on the soles of my feet or running their fingers through my hair to compare its texture with that of their own. Usually I did not object too much. The Shavante do not make so much of physical contacts as we do and I did not want to appear fussy. But on this particular day it was very hot and sticky in between rainstorms and I was perhaps over irritable. I seemed to be half buried under these three young amazons and I could not even send them away with the excuse that I had weighty matters to discuss, for the man of the house had gone outside and I was just lying there hoping he would return. I pushed the biggest of the girls away, adding rudely 'What do you seek?'

which is the Shavante way of putting a relationship on a formal footing. She laughed and her sister jumped up in an exaggerated pantomime of fear.

'*Pipa-di,*' she said, cowering away. 'I'm afraid of you. You're a spirit. Go away, *wazepari'wa.*' With emphatic motions she banished me to the far horizon and then burst out laughing.

At that moment Domitsu'a, the idiot, came into the hut. He looked at me, grimacing, and I did not greet him. He was a kleptomaniac and had, among other things, stolen Pia's sheath knife, a treasured souvenir from Lapland. Everybody knew he had stolen it but he refused to give it back, and so we had made it clear that he was unwelcome in our house unless and until the knife was returned. But at that moment I was not thinking of the lost hunting knife, I was watching the dilated eyes of the girl who had been teasing me seconds previously and noting the small clenched fist she had raised to her trembling lips.

Domitsu'a went out, and I turned to the terrified girl.

'Why are you frightened?' I asked.

She did not answer, but stood looking at me wide-eyed and still gnawing her knuckles. I turned to her elder sisters who were obviously shaken too.

'The spirits,' they muttered, pointing to the man who had just gone out, 'he talks to the spirits.'

Then I recollected that Domitsu'a slept in the next house and I guessed why the men in this part of the village were so anxious to get torches from me. Domitsu'a talked in his sleep and probably had fits too. He would start up suddenly, his features twisted, babbling incoherently, and his fearful neighbours must have assumed that he was in communication with the sinister *wazepari'wa.* Now I understood why Domitsu'a and his vagaries were tolerated by his fellow Shavante. All of them entered into contact with the spirits at some time or other when it was necessary to do so, but none of them sought out the *wazepari'wa* to commune with. A man who did such a thing was like a lightning conductor—imbued with dangerous power.

Ordinary Shavante aspired to such power only at certain times and then only under very special conditions and with the most solemn rituals. They called this ceremony the *wai'a,* and during the days of its performance all women and uninitiated males were confined to their huts. The men sang all day in the jungle and communed with the spirits, that they might receive power. Sometimes it was the power to hunt successfully, and at such a *wai'a* the arrow canes were blessed for the good of the community. Sometimes it was the power of life and at others the power of death.

We had heard much about the *wai'a*, how the women went in terror of it and even fled to the mission for sanctuary when the men at Sta Terezinha were performing it, how the life of the community depended on it, how it was the most important ceremony in a man's life—the goal for which his separation in the bachelors' hut and his initiation had been mere preparations. We were, therefore, especially excited when we heard that it would be performed in the near future for the new initiates.

I heard about it in the men's circle and told Pia later that night; but she had already had the news. The women, she said, had been whispering about the *wai'a*. The whole village was talking of it. Indeed in the days that followed they talked of little else. The men hunted assiduously to collect a surplus of meat for the ceremonies. The women did what they could to seem uninterested in what was going on. But there was excitement in the air, especially among the young men. They sat self-importantly in our hut and told us about the *wai'a* and how the women were afraid of it. They might not see it on pain of the most terrible consequences, we were informed, and the directors of the ceremony went round the huts to make sure that they did in fact stay inside and not desecrate the mysteries. One young warrior who had always been very friendly with Pia made her a pair of white wristlets from the inner bark of a certain tree and urged her to wear them.

'For the *wai'a*,' he insisted. 'They are beautiful. Everybody wears them for the *wai'a*.'

'Everybody?' Pia asked him. 'But women do not take part in the *wai'a*.'

'Everybody,' he told her urgently. 'Wear them! They are very beautiful. Other women wear them too.'

Long before dawn the older men were gathered in the middle of the village, chanting. It was a beautiful, surging rhythm and I tried hard to remember what it was it reminded me of as I wandered out into the grey vestiges of night. I set a course for the fire which flickered invitingly where they sat. As I came up to the group they finished their song and broke into a deep breathless bellow.

'*U, he, he, he . . . U, he, he, he . . .*'

I suddenly remembered the Sherente communing with the

spirits of their ancestors. They had sung in the same fashion and it was the same call that Wakuke told me the powerful shamans used in the old days before the white man came.

A party from my age-set were parading round the houses now and I went with them. They entered each one unceremoniously, stamping and snorting like angry buffalo until everybody was awake. If there were young men present who were slow in getting up, they seized them and dragged them out of the huts, picked up their sleeping mats and hurled them after them. By sunrise the huts were cleared of initiated men, and ceremonial leaders were patrolling the village to ensure that all door and window openings were properly sealed with thatch. Then the men retired to the forest and stayed there chanting.

It was well past midday when we were ready to enter the village again. Everybody was in his full paint, his hair freshly tonsured and oiled, his body shiny. The procession was unusually solemn. Without any of the laughing and horseplay that usually accompany Shavante rites all the men paraded into the village. Two men carried the sacred arrows and two others carried a stylized representation of a human phallus. These were ceremonially danced around the village and even taken into the huts, where the terrified women chased their bearers out again.

I went back to the hut and found Pia impressed despite herself. All day she and Biorn had been cooped up in the stuffy interior while the chanting throbbed around them. She had peeped out occasionally and seen men coming and going in their full regalia, solemn as proconsuls, till she could almost feel the suspense building up in the village.

'Imagine what it must be like for the other women,' I said.

She asked me the significance of the white wristlets and anklets which she was wearing.

'I don't know,' I told her. 'All the men put them on and I assume it is the correct uniform for the *wai'a*.'

'But have you seen any women wearing them?'

'I haven't seen any women in the last day or so,' I confessed, 'anyway not enough to notice whether they were wearing anklets or not.'

'Well, I talked to some today,' she told me, 'and they were horrified when they saw me wearing them. It's for the *wai'a*—they kept telling me and pointed to another girl who had got them on. I only saw one other girl with them on in the whole place.'

'I should take them off,' I advised her hastily.

Next day the men retired to the forests once more and again their chanting pulsated through the village, mingled this time with the mournful note of a hollow gourd hooting in the trees.

'Listen to the spirits,' the men told me, and sang even louder. Four of the elders beat time with their dance rattles, and in the middle of it all Apewen sat as still as a statue, pondering the spirits of death.

Now the young men were brought in, those who had never before attended a *wai'a*. They stood with their eyes fixed on the ground, quivering slightly but never recoiling as their elders danced at them. It was a dance fit for Mars; thundering, snarling, vicious, and in that setting, terrible. The dancers twisted their faces into masks of bellicosity, lifted their knees to their chins and stamped the earth like berserks. They snorted and uttered hoarse grunts of rage as they advanced and recoiled and advanced again, attacking the initiates so closely in the dance that they sometimes stamped on the arches of their feet. But the initiates remained, motionless as Apewen, till the performance was over.

In the afternoon we filed back to the village again and danced all round the houses where we could hear the women and children scuffling like frightened mice. Then Apewen took me to the centre of the village and the old men gathered there. I guessed that I was being tactfully removed from the climax of the ceremonies, and a moment later I knew I was right. The celebrants all filed away into the forest. Two of them only remained behind and went into one of the huts. A moment later they came out with a young married woman. I could see from where I was that she was very frightened, though she was trying not to show it. One before her and one behind they marched her into the forest. She was wearing white bands round her wrists and ankles. The men returned and led another woman into the jungle and then a third, so that every clan in the village had given up a woman. All of them wore the white bands of the *wai'a*. Out in the forest they were forced to have intercourse with the sponsors of the ceremony, with the warriors and finally with the newly initiated boys.

That night there was dancing and jubilation in the village. The men had interceded with the spirits and received the power of life and the power of death, sexual potency and strength against adversaries, both of which were symbolically demonstrated in the culminating ritual of man against woman. The *wai'a* was over.

Yet, unknown to me, the Shavante felt awkward about the way they had treated me. I was, after all, a member of an age-set, I had danced with them and accompanied them through all the ceremonies of the initiation cycle. I had attended the men's council as my age and status demanded, yet, when it came to the crucial stages of the *wai'a* I had been treated as an uninitiated boy. They decided to make amends.

I was standing watching the dancing when Sibupa suddenly lunged at me. I was so startled that I jumped back about a yard which was, of course, quite the wrong thing to do. He came at me again, snorting and stamping, his face twisted with fury, and I realized he was dancing at me. I stared at the ground and did my best to keep still. But Sibupa was working himself into a frenzy. He grunted in my face and his flailing elbows almost knocked me backwards. He stamped on my toe and then seized my head between his hands and bit me on both cheeks. Here I would have recoiled instinctively—if I could, but Sibupa held me firm for a moment before he stepped back into the circle of elders.

I stood utterly bewildered until Apewen stepped forward and made a speech. I was a true *wai'a* celebrant, he informed me. I was a true *wai'a* celebrant . . . he repeated himself again and again as Shavante do, and anyway I could not follow everything he said for he spoke in the formal manner, and the blood was beginning to trickle down my cheek. They sent for a pair of scissors and a woman came over to tonsure my hair in the firelight. Then they painted my crown red with *urucu* in the proper ceremonial fashion.

It was late that night when I got back to the hut, and Pia was already in her hammock. Mine hung knee-high between me and the box on which our candle stood. I stepped over it, caught my foot, shook it loose and in the process fell full length with a crash that echoed inside my head as if I had fallen through a corrugated iron roof. I had in fact caught my forehead on the sharp edge of a tin of dehydrated vegetables. By the time I got the lamp alight the hut was already filling with Shavante who had come to see what happened.

Pia looked at me, at the lump rising like dough above my eye, at the marks on my cheeks, at my painted crown.

'Good God,' she said weakly.

Pia and Biorn were due to leave. The commandant at Campo Grande had promised to send an aeroplane for them after a month or two so that they would get back to the coast in time to take ship for Denmark. We were vaguely aware that unless the plane came soon they were liable to miss their boat, but the whole idea of sailings to Denmark was phantasmagorical besides the realities of life among the Shavante. Especially now that Apewen and all his household had moved in with us.

They did not tell us about it or think of asking us. One day Suwapte started taking the thatch off their hut, and when we returned to our house we found hordes of young men sticking their bows in our ceiling and looking for likely places to hang their baskets. We could hardly complain since by Shavante standards they were such close relatives of ours, but we found their visit—as so often happens when relatives arrive unexpectedly—a mixed blessing. It taxed our floor space to the limit and made it almost impossible for anybody else in the village to come and see us, for they could find nowhere to sit. And then the comings and goings that went on all day and all night made our hut feel like army headquarters on the eve of an important advance. Unlike the popular misconception about Red Indians, Shavante are not delicate in their movements. They are thunderously flat-footed when they run, and they barge rather than slip in and out of the houses they visit. Each time they barged into ours they knocked against our hammock ropes, so that sleeping there was like being adrift in an open boat on a heavy sea.

After the first night, Pia and I gave up the attempt to sleep well before sunrise and prayed that our guests would soon go about their daily business, just to empty the hut a bit. It rained, though. First the dank, drizzly, early morning kind of rain best calculated to keep even the Shavante indoors, followed by a grumbling thunderstorm which battened on the day without actually exploding and getting the rain out of its system. We found ourselves cooped up with fourteen men, eight women, several children, a bitch and seven blind puppies. I am not sure of the totals for our fitful candle did not light up the corners of the hut, and anyway the throng was as restless as a snake pit and it was very difficult to count them.

I am clumsy enough as a rule and when we were alone in our hut I usually contrived to step in pans of food or knock my head on protruding branches, even when I was not spectacularly engaged in diving into vegetable tins. During these days, though, the house perpetually resembled the scene of a custard pie comedy. Whenever I moved the dog growled at me and Pia gave me admonitions not to knock over this or that. At intervals the children upset the tape recorder or tumbled off our impromptu shelves, bringing their contents cascading down with them. Our lamp was knocked over so many times that we finally gave up the effort to use it and simply lay there in the twilight, ignoring the spilled pots and the yelps of the children who had got under somebody's feet. It was impossible to carry on a coherent conversation with anybody under the circumstances, so I could not even employ the time to gossip with Apewen.

Biorn enjoyed himself. He was learning how to dance, and so he shuffled round the hut with Suwapte's little son, imitating the Shavante songs in his squeaky voice while the puppies he trod on added to the chorus. Sometimes he would scuffle with the other children for bits of food under our shelves, but luckily he rarely won. It was usually the tailless macaw (its prized tail feathers had long ago gone into ceremonial regalia) who came in and snapped up the booty while the children were still wrestling. Most of all he was enjoying his comparative freedom from parental interference. The Shavante disapproved strongly if we disciplined him, even if it was for taking advantage of their children. Men sometimes sought me out at the river or in another house to tell me in shocked detail how his mother had spanked him. So we cut down on the spankings and let him run riot, an arrangement which appeared to satisfy everybody except us, who found the tumult in the hut intolerable.

On the second day the weather cleared. With the better visibility came the aeroplane from Campo Grande and within the hour Pia and Biorn were spirited away. One moment we were on our way down to the creek to bathe, and the next we were in a hectic throng of Shavante, running helter-skelter to the airstrip with Pia's trunk and all the bits and pieces she was taking back with her. Biorn stoutly refused to put on any clothes, and I remember him trotting stark naked towards the plane with his little red travelling satchel in his hand. A few seconds later I was straining for a glimpse of Pia as she settled herself in the aircraft, and then I was left with the Shavante and only that hum in the empty sky to remind me that I had not dreamed it all.

The Shavante certainly refused to believe that their departure was final.

'They will return,' they kept reassuring me in the tone of voice one uses to a very sick patient. 'With the next aeroplane they will come.'

Apewen was the most insistent of them all. He felt it in his belly, he told me, that Sibupa would soon return. I was beginning to wish he was right. Another two days of the enforced company of all Apewen's kinsmen had made me feel very lonely indeed.

At last the rain held off long enough for Suwapte to complete the scaffolding of the new hut and start the women on the thatching of it. That night it was only half finished, but he and his wife and children moved into it anyway just to get away from their in-laws, and their fire glowed eerily through the open roof, making the whole structure look as if it were floodlit. At least, I thought, I had heard no more of the sorcery charge which was brought against him. Perhaps the villagers had been too busy. Now, with the *wai'a* over and their houses rebuilt for the rains, they could take up their feuds where they left off.

By now I could understand the pattern of these hostilities. Each Shavante community seemed to be sharply divided into two factions, the dominant group, consisting of the chief's clan and its supporters, versus the rest, the outsiders. Of course the membership in these factions shifted as groups of men took up different alignments according to their understanding of their own best interests, but the structure of the situation remained the same. The chief maintained his position via the fact of being supported by the strongest clan, and the Shavante thought of his faction as being in formal opposition to the outsiders in the community. When two men faced each other as speakers in the men's council, they were invariably from opposite factions. It was in fact a sort of two-party system, but it went much deeper than that. The idea of it was embedded in Shavante speech. A Shavante conventionally assigned any other member of the tribe to one or other relationship category, thereby making him a 'relative'. But he did this by a sort of rule of thumb which classed people of his own clan under the categories reserved for 'kin' and people of all other clans under the categories reserved for 'non-kin' or 'affines'. It was from his kin that a man expected solidarity, from his affines that he feared the worst. Hence the ambivalent attitude of young men after marriage. To admit to being married was an admission of their inferior status in the house of their affines, a house where they worked for their father-in-law for the benefit of their brothers-in-law. Indeed it was in the household that proximity between kin and affine bred the most tension. That was why 'brothers-in-law were always angry' and why a man looked first to the affines married into his own house if he suspected anyone of trying to bewitch him.

Indeed Shavante appeared to regard a state of permanent tension between the dominant faction and the rest, between kin and affine as the natural order of things. Even their cosmology reflected this preoccupation. Spirits of dead kin, they told me, went to the village of the dead which was a place of happiness and abundance which lay at the beginning of the sky to the east. At the end of the sky to the west lived the sinister *wazepari'wa* spirits who lay in wait for the souls of the dead as they made their perilous way eastwards and tried to carry them off to swell their own ranks. When I had asked how the village of the dead to the east could be so idyllic when everybody's kin went there . . . and therefore everybody's affines too, the Shavante had replied in some embarrassment that in the village of the dead there was no fighting and no sorcery. Since fighting and sorcery were the characteristic modes of relationship between kin and affine, I concluded that the village of the dead was essentially a 'kin' place where affinity had been exorcised. On the other hand the village of the *wazepari'wa* was in every way its antithesis, and *wazepari'wa* actually presided over the marriage ceremony! So the opposition between 'kin' and 'affine', between 'we' and 'they' ran right through the Shavante scheme of things, ordering their way of thinking and their way of acting towards one another.

But sorcery was currently a less urgent problem than jaguars. The rains had not dispersed those which roamed in the neighbourhood of the village, possibly because the cattle belonging to the post were too strong an attraction. Each night they could be heard calling to each other down by the river, at the creek or around the village itself. Women did not dare to stir out of their huts after dark, and a number of children had been badly frightened.

Parties of men began to go out after them, and I went too, for I was curious to see whether a jaguar hunt was different from any other sort of hunt as far as the Shavante were concerned. The fathers at Sta Terezinha had told me that they wore specially reddened earplugs and performed certain rites in order to ensure their own safety and the success of their hunt when they went after jaguar; but I saw nothing of the kind. Shavante seem to be the most unceremonious people, except in the *wai'a*, which is probably why that performance was so awe-inspiring. When we looked for jaguar we were simply more single-minded than when we hunted other game. Even so it was difficult for my companions to resist the temptation to turn aside if we came across fresh tracks of tapir or peccary. Sometimes they did not resist it, which was one of the reasons why we did not flush the jaguars out of their retreats very quickly. Once I travelled with a party whose members were wild with excitement at the nearness of the great *hu* we were trailing, but we lost him at nightfall and even Shavante are unenthusiastic night trackers when they are out after jaguar.

One day we hit a fresh trail in the early morning and my companions felt their prey was in their grasp. I had no particular impression of being hard on the heels of anything, for I had long since accepted the fact that patience is more useful than élan when one is hunting with Shavante. I simply did my best to keep my .222 from jarring against my pelvis as I walked, without any particular hope of seeing a jaguar. In fact, I rather hoped we did not meet one—at any rate too suddenly. On the other hand, I knew that we would be out all day unless we did and that was not a very pleasing prospect either. I was shuffling along, trying to calculate the probability of rain before nightfall, when my companions bounded forward and everybody seemed to be hopping with excitement. I unslung my rifle before I heard the snarl of the jaguar. Everybody was shouting 'Shoot'—'Quick'—'Kill it' and a medley of other injunctions which I did not realize were addressed to me until I noticed that I seemed to be the only gun in the party. The others were fitting arrows into their bows, and I was wondering where the other rifles had got to. I had not yet had a proper glimpse of our quarry. He seemed to be a long way away for my shooting capabilities and yet . . . how long would it take him to cover the distance if I only wounded him? Running for a tree would not help. Jaguars climb better than men do. He was walking about, snarling and swishing his tail. I wished he would keep still.

'*Bare-e-e-e-na*,' the men squealed, imploring me to hurry. The next thing I knew I had fired. The jaguar whisked round as if to reprimand me and let out a roar which, as the saying goes, would have curdled my blood had the issue simply been between him and me. But it was not. I was taking a public examination in hunting skill before a group of Shavante warriors and it was a subject in which everybody expected me to fail. I found myself firing more cold-bloodedly than I would have given myself credit for, ignoring the beast and its antics entirely. I thought it was enraged. I thought it was preparing to spring. I thought it was running away. All these thoughts were quite separate from the part of my mind which was thinking: 'They are only ·22 bullets. One may not kill him. Keep firing. Perhaps their arrows are more effective anyway at close quarters.' The animal was doing none of these things. It was rolling in its death agonies and the Shavante were closing in on it, whooping and jabbering in their excitement.

We got back to the village early, and just before sunset Tenente Haroldo breezed in again in his miraculous plane, bringing letters from Pia and parcels of clothes which she had sent up for the Shavante. They were delighted. The men especially felt it had been a wonderful day. They sat in the council fingering their gifts and telling me over and over again how they missed Arenwain'on and how they hoped she would return to them soon. Wearily I told them once again that Arenwain'on was going back to her country which was not my country. That it lay far away across the great water in a place which even aeroplanes took many sleeps to reach. They listened earnestly and when I had finished they told me that she would soon come back.

I should have felt elated, I suppose. I had shot a jaguar; its tawny skin on my living-room floor would be a sure passport to that band of bores, the old hands who knew the country. Pia and I and even little Biorn had lived with some of the wildest Indians in Mato Grosso and even come to like them after a fashion. Now these same notorious savages were apparently anxious that Pia and Biorn should come back to them. My research too had been successful. I had set out to study a moiety system and discovered that the Shavante saw their own society as a thoroughgoing system of opposites, yet dispensed with the institution of intermarrying moieties. This would put a different complexion on the comparative research I hoped to pursue both in Central Brazil and in other parts of the world. Surely now was the time when I should have felt most at ease, both with the Shavante and with myself. Instead I felt utterly worn out. I did not cherish the pathetic illusion that the Shavante accepted me as one of them or my family either. They tolerated us. They might even be happy to see us come back, provided we brought plenty of presents. But they could not speak freely with us. Even if they could, we were separated by a barrier to further understanding which I wondered if years of field work could penetrate. The task appeared utterly hopeless. At that moment I saw the whole world atomized, each individual wrapped in the impenetrable cocoon of his individuality, inscrutable. People could not understand people. I had known this as a constant theme in European literature. What was worse, when I was tired I believed it, so what did I think I was doing among the Shavante? I decided I needed a rest.

But there could be no rest until the commandant sent a plane from Campo Grande for me. I had to stay with the Shavante, still wondering, puzzling, noting and recording; still teased by doubts and lonelier than ever before. The Shavante were preparing to go on trek. I would not go with them. My life was already diverging from theirs and yet it was still bound up with theirs as it had been for years.

When the plane did come I felt an unaccountable sense of loss, as if my own personal impetus were exhausted and I was adrift, purposeless. Apewen embraced me and made a speech. We must come back, he said. We must come back. Arenwain'on must come back. Sibupa must come back. We must bring many presents. We must bring many fish hooks. We must bring many balls of ammunition. We must bring many clothes. Yes, already they missed us. We must come back.

I looked at the old fox, trying to follow his rhetoric, and it was then that I noticed he had tears in his eyes. Perhaps, after all, he really meant it.

Epilogue

When we went back to the Shavante, we could reach them by road. It was no longer necessary to beg (and wait) for rides from the Brazilian Air Force. We could pack all our gear into a Volkswagen bus and drive to the region that had seemed so mythically remote to us a quarter of a century earlier.

But we still had to deal with the Brazilian bureaucracy. Since anthropologists, both Brazilian and foreign, regularly denounced the government's treatment of the Indians, the government made it as difficult as possible for them to get to the Indians at all. Ranchers, prospectors, journalists and even tourists could get to Indian country more easily than anthropologists, who were strictly controlled. Foreign anthropologists had to present elaborate documentation months before coming to Brazil, including chest X-rays to insure that we at least would not infect the Indians. These had to be presented so long in advance that they were useless as guarantees of our being free of infection in any case. Approval of this documentation authorised the anthropologist to proceed and to receive the appropriate visa, but did not guarantee access to the Indians. That still depended on FUNAI, the Indian service.

FUNAI had perfected a subtle way of discouraging anthropologists. Weren't we the people who were always insisting that the Indians ought to be consulted, that they ought to have a say in their own affairs? Well then, FUNAI, who rarely asked the Indians about anything, insisted on elaborate consultations to discover whether Indians were willing to receive anthropological researchers. This would have been perfectly all right except for the time it took and for the fact that we all knew it was not the Indians themselves but FUNAI agents at Indian posts, or missionaries or other intermediaries who answered these requests. We were informed, for example, that the Shavante at Pimentel Barbosa, the village we knew best, were reluctant to have us return. This in spite of the fact that they had sent invitations to us through intermediaries and old Apewen had recorded a tape for us, telling

267

us that he missed us. Apewen, alas, was dead now, but we doubted that the sentiments of his people had changed so abruptly.

Our Brazilian colleagues interceded for us of course, but I think it was the traumatic effect of our son's illness as much as anything else that finally got us our permits. The FUNAI officials were moved, despite themselves, as we explained that my son had contracted cancer and returned to the United States, so that permits would no longer be necessary for him and his wife; so they issued them to our depleted expedition and assured us that they would give us every assistance if we wanted to come back again and finish our work on a more auspicious occasion.

A few days later we reached Barra do Garças, the boom town that had grown up on the opposite bank of the river from Aragarças where our long river journey with baby Biorn had ended. The town, indeed the region, had clearly been taken over by southerners, gauchos from the state of Rio Grande do Sul, seeking land in the vastness of the interior and hoping to get rich on cattle and rice. The town's eating places served southern-style barbecues and played lilting gaucho tunes, very different from the melancholy country music of the locals. The locals were not in any case much in evidence and the local Indians even less so.

Only the countryside seemed unchanged. The monotonous scrub, interspersed with jungle wherever there was enough water and occasionally soaring into spectacular buttes and mesas reminded us powerfully of our days with the Shavante. But where were the Shavante?

The village of Pimentel Barbosa had moved away from the Rio das Mortes and could now be reached by driving through a miserable little hamlet called Matinha and then turning off the dirt road onto the rutted trails that led through Indian country. But the trails branched and forked at will and we followed them by guesswork, for there was nobody from whom we could ask the way. We made detours to huts or occasional farmhouses to ask directions but found them deserted.

Barra do Garças was abuzz with stories of how bands of Shavante had come to these homesteads one by one and forced their owners to saddle up or get into their Toyota trucks and leave there and then. Weeping women would describe how they looked back for the last time at their homes and saw Shavante going through them and carting off their possessions. These were the people who had moved onto Indian territory, secure in the knowledge that Indians, once they were 'pacified,' never managed to evict *civilizados*. Only they had lost their gamble and (they forgot to add) had ignored the Shavante's repeated warnings to get out before they were thrown out.

We drove on past these ghostly, deserted houses, watching the sun sink low in the sky and wondering what we would find at the end of the trail. Suddenly they were there. The trail came over the crest of a hill and plunged down a steep slope. We looked down through a gap in the trees to an open plain and saw, laid out like our memories far below us, the horseshoe sweep of huts marking a traditional Shavante village.

It was half an hour before we got down to them. The trail led first to the squalid houses of the Indian service. It was there, as we presented our credentials, that we met Sibupa, Apewen's son and the Shavante namesake of our own son. Sibupa embraced us and immediately made a speech of welcome before a gaggle of wide-eyed youths who barely knew who we were. We were taken in procession into the village and seated in a shelter in front of the chief's hut, where we were soon suffocated by the greetings of old friends and the curiosity of the younger generation. Waarodi, now the chief after the death of his father Apewen, appeared, sat down and began to interrogate us in the formal style.

Where, he asked, was our own son Sibupa? How big was he? Was he married? Did he have children? Why had he not come back with us? We answered his questions, close to tears ourselves from the emotions of the moment and from having to say out loud that he was very ill and we did not know if he would get better.

That night I was put through it all again in the men's circle. The men of my age-set welcomed me, patting their mats invitingly for me to sit next to them, but then I got a nasty shock. In the old days I was allowed to sit and listen. Now I was an elder and I was expected to speak. I was urged to my feet and obliged to give the most embarrassing speech of my life. Shavante are fine orators and connoisseurs of good speaking. It was especially humiliating then to have to stand up before them and stumble through what I could remember of a language I had not spoken for fifteen years. It was a relief to get through it and to hear a kind friend exclaiming, with more charity than truth, '*He wa t'prin*'—the Shavante version of 'Well done!'

My age-mates asked me if I had retired. I told them that I certainly had not. I might be grey-haired, but was still active enough. 'Oh but you should,' they told me. 'We all have. It is good to retire. Then you get money.' They had discovered that they were entitled to social security payments and now, on the basis of claiming to be about the right age, were all collecting about eight dollars a month which, since they had never had any money, they considered quite a windfall.

There were other changes too. The young men's age-set were all away working in the rice fields. Only older men and youths were left in the village. Many of the teenagers could read and write and they played soccer so enthusiastically in the evenings that the men's council could not meet till after dark when the center of the village was vacated by the players.

They asked us constantly for rides, most of which we had to refuse, for it took all day and a lot of gasoline to get anywhere in those parts. Besides there was the Indian agent and his truck. This agent lived in the filthiest house imaginable, which did not bother him much because he spent all of his time either repairing his vehicle or driving around the country in it. He obviously had not much idea of what he was supposed to do either with or for the Shavante and we could easily understand why he had earlier sent word to Brasilia that the Shavante did not want us in their village.

We did go on one long excursion though. Surupredu took us to the rice fields to bring home the young men so that they could take part in a ceremony. Surupredu had grown into a person of influence, wise in the ways of the outside world. He dressed smartly in army fatigues and a checked cap, spoke good Portuguese and had travelled much throughout Brazil. His son was going to school, through the courtesy of friends of his in the state of São Paulo. Yet, in his own way he was a traditionalist. He was often the main speaker in the men's council and the leader in traditional ceremonies. I discovered that Pahiri'wa, who had led me into the Snoring Mountains so many years ago, had now broken away from Pimentel Barbosa with his own group of followers and founded a separate village. I wondered how long it would be before Surupredu did the same. He did it the year after our visit.

Waarodi, the chief, never appeared in the men's council during our stay. He was ill and was keeping to his hut. In fact it was for him and for our own son that the villagers decided to perform a curing ceremony, the *wai'a* for the sick. That was why Pia set out with Surupredu to bring back the young men. Pia was quickly becoming our designated driver, for she did not mind taking the Shavante on errands while I got on with the work of talking to the ones that were left. On this occasion she discovered that Surupredu had not lost his hunter's eye, for all of a sudden he ordered her to stop the car. 'Sit still!' he commanded as she braked it to a halt. Then he leaned across her, stuck his long-barreled pistol out of the window and killed a deer with a single shot. It was a jubilant group of young men that came back, deer and all, in the overloaded bus that evening. We never did manage to get the deer's blood out of the vehicle after that.

Meanwhile thunder clouds were gathering as if to match the serious mood of the village as the men were preparing for the wai'a. The elders had planned it in whispers the night before in the men's council. The whispering was not for fear of being overheard but because these are the tones appropriate for the celebration of such a solemn ceremony. All day the men stayed in the forest a little way from the village, singing and painting themselves. It takes enormous care and patience to paint one's body red and black, using only one's own hands to apply the charcoal and crushed urucu seeds, to get the dividing lines between the paint surfaces clean and clear, and above all not to smudge the fresh white palm strips that have to be bound round both wrists and ankles. When they were finished and were binding the tall headdresses of macaw's feathers into their long hair, the men looked exactly as they might have done when I first met their parents years before, except that now they all wore scarlet soccer shorts to complete their regalia.

As the sun sank low in the sky I took my place with the elders who were seated outside the chief's hut. The singing ceased and even the dogs and children seemed hushed as the men filed into the village. They walked solemnly round the great arc of their world, about seventy of them, from middle-aged men to lads barely in their teens while the thunder rumbled among the vermilion clouds. Then they formed a tight semi-circle in front of the chief's hut and started to sing and dance.

Now Waarodi appeared at the door of his hut, incongruous in bright blue pajamas. He lay down on a mat and let the singing wash over him. I could see Surupredu at one end of the line, painted like the rest, his tall feathers soaring and swooping as he bent in the dance, thrusting forward the two white wands he carried and willing the sickness away. Night came down but we knew that the singers would stay there till dawn.

271

They were soon hoarse. Hours of singing in the chill night of the Central Brazilian plateau saw to that. Yet they went on croaking and dancing, occasionally joined by a fresh voice that had taken a rest and come back into the lead. Soon we too lost all sense of time. There was only the singing, like a force of nature come to do battle with disease. Shadowy figures sought us out to make sure we understood. 'It is for Sibupa as well as Waarodi. Your son will be cured.' Pia went to her hammock and wept.

We did in fact get news of Biorn when we left Pimentel Barbosa and returned to Shavantina, for that little town had just acquired its first telephone. We went down to the phone company's office and put through Shavantina's first international call. The local people, who were thronging the office as they waited patiently for their calls to other cities to come through, were fascinated. They crowded round Pia, listening to the foreign language and commenting on the marvel of speaking to the United States, while she tried to learn whether Biorn had started his treatment and what the prognosis was. He did get better, but all the time we were with the Shavante his illness hung like a pall over us. The worst of it all was that we were travelling from village to village and at every community we had to explain all over again why he was not with us, that he was very sick, that we did not know whether he would recover. . . .

As we travelled we realized that Pimentel Barbosa was one of the more conservative Shavante communities. They kept their village a day's journey away from their rice plantations and equally far from the delights, such as they were, of Matinha with its flyblown stores and two bars. In fact they had recently rebuilt their huts in the traditional beehive shape, after experimenting for a time with rectangular Brazilian houses.

At Ariões we found a village still built on the horseshoe plan, but whose houses were mostly Brazilian-style huts. Here the most energetic Indian agent we had ever seen presided over the most sophisticated farming. There was a huge hangar full of tractors and farm machinery and silos full of rice being harvested from fields only five minutes' walk away. We were now accustomed to the sight of Shavante driving tractors, but we were fascinated by the ambition of the enterprise at Ariões. This was the most dramatic example of the Shavante Project that we had heard so much about. Were the Shavante really being taught to become self-sufficient as tractor-driven rice farmers? The answer, unfortunately, was no. The capital, the machinery, the management and most of the skills were provided by the Indian service. The Shavante swaggered about and spoke of it as their project, with pride of ownership in their voices, but I doubt whether they really believed it. They were in truth the hired hands on their own enterprise and they understood their situation better than they admitted. They cross-questioned us about whether the Project would be funded again next year, whether it would be funded at the same level, or—with real anxiety—whether it might not be funded at all. But they consoled themselves with the thought that if the government faltered in its support of them, they would once more march on Brasilia and make the authorities change their minds. Those were the days when the Shavante were flushed with the success of their own truculent lobbying. Now that Brazil is in the grip of a major economic crisis, the Shavante project is fading and the Indians are desperately casting around for some other way of entering the national economy.

Such gloomy thoughts were not much on the minds of the Shavante of Ariões when we saw them in 1982. They had only recently had their lands guaranteed and they were learning how to cope with the dramatic changes that were altering their lives. Now that they lived in and for the Project, they had virtually given up their traditional wanderings and their village was starting to acquire the look of a settled community. There were houses built behind houses in the village horseshoe and chicken coops and pigsties springing up everywhere. Worst of all, as so often happens when nomadic people first settle down, the village was silting up with refuse. Previously the Shavante had simply moved away from their garbage and traditional Shavante villages were kept scrupulously cleared. Now the sign of civilization seemed to be the accumulations of trash.

But it was at the Salesian mission at São Marcos that we found the most 'civilized' Shavante and the experience was unnerving. Indeed our whole stay at the mission was unnerving. It is set in the wildest and most dramatic country that we had seen. We took the wrong track on the way there and barely escaped being bogged down for good in the sand. On our arrival we pulled into a courtyard surrounded by mission buildings and were ignored by everyone. Finally a Shavante to whom I introduced myself pointed out Adalbert Heide, the lay brother who had helped me with my Shavante texts at Sta Terezinha years before. I had acknowledged his help gratefully in my book *Akwẽ-Shavante Society* and so I went over to him expecting some sort of welcome. I had to tell

273

him who I was, for the intervening years had made us both un-recognizable to each other, and he fetched the father superior. That was the last time he spoke to us, though we spent a week at the mission.

The father superior asked us who we were and I explained that I was the anthropologist who had worked with the Shavante many years ago, that I had returned to do more research among them, and I introduced my wife and my friend the photographer. To this I received the startling reply that yes, he did remember being consulted by FUNAI about an anthropologist who wanted to return to the Shavante, but he had replied that the mission did not wish to receive him. At least he made no pretense that it was the Indians that did not want us to come. I produced my authorization to work at São Marcos and finally, with obvious reluctance, he told us we could stay.

We were housed in comparative comfort in rooms giving off the stone-flagged veranda on one side of the mission. We were summoned at meal times to a room where food was set out for us by the mission staff who then withdrew and left the three of us to it. In effect we were in limbo, kept at arm's length by the mission, who treated us as if we were infectious, and yet not living with the Indians either.

Nor was my initial contact with the Shavante too promising. I learned that the chief of this village was Aniceto, whom I had known as a teen-ager back in Sta Terezinha. I had thought of him then as a typical mission boy, always hanging around the priests. They had taught him to read and write and I had paid him to help me with my Shavante texts. Now he appeared as a self-important middle-aged man who wasted no time reminiscing about the old days. He summoned me into one of the mission classrooms and demanded to see my credentials and my author-ization to work at the mission. I showed them to him. He cross-questioned me about the research I was doing and I explained it to him. Then he lectured me about the perfidy of white men and the benefits that his people had acquired from the missionaries; after which he ignored me.

It gradually became clear that the Salesians were exceedingly defensive about their work. Their missions in the Rio Negro area to the far north had been attacked for exercising a virtual he-gemony over the entire region, and their missions elsewhere crit-icized for being too authoritarian. The missionaries therefore felt persecuted and misunderstood. They were reluctant to have anything to do with people, such as anthropologists, whom they believed would only criticize them and this paranoia had certainly been communicated to the Indians under their charge.

We quickly discovered the extent of their control over the Shavante in their care. São Marcos, the community attached to the mission, was the largest Shavante village—it was virtually a town of about a thousand inhabitants—that we had ever seen. The mission also looked after three other villages that were visited at intervals by fathers and nursing sisters. Mario Juruna, the Shavante congressman, told me that he had grown up in one of these satellite villages and had left it to start his career among the Brazilians because he could not stand the control of the missionaries.

The first night that I ventured into the men's circle at São Marcos I noticed that something, I could not quite say what, was different about it. Then I realized that it was after dark and yet the men who had gathered there were plaiting sleeping mats and making other things as they talked. This could not happen in a traditional Shavante village, where men concentrated on what was being said because there was no light to work by. Here the center of the village was floodlit! I looked up to trace the source of the light and saw that it came from a huge illuminated cross that towered over the community. Soon a voice bellowed admonitions and instructions from the direction of the mission. Traditional Shavante villages face (i.e., have the open end of the horseshoe) towards their source of water. This village faced towards the mission church, from which the missionaries could harangue the Indians over a loudspeaker that brought the voice of their guardians into every corner of the community. I began to understand Mario Juruna's rage. Living in a community like this would be like being sent to boarding school for life.

Yet it was also true that the Indians at São Marcos probably had the best schooling—albeit with a strong missionary slant—and the best medical care of any Shavante. We had plenty of opportunity to sit in on classes, for that was one of the major activities of the mission. We also had a chance to hear the classes that Aniceto was conducting for his people on how to vote.

Indians are considered minors under Brazilian law. They are formally wards of the state, with FUNAI as their tutor. The Indian service appeals to this legal status when it tries to prevent Indians from travelling in Brazil without permission or from holding meetings of which the service disapproves. Yet this formal provision is otherwise largely ignored. FUNAI is a most derelict guardian and Indians are permitted to exercise other rights of citizenship, like voting and being elected to political office. In 1982 Brazil was holding its first free election since the military had taken over eighteen years previously. The military government that was still in power had done everything possible to frame the election rules in such a way that their party would not be humiliated at the polls. One such provision was that the ballot was long, with candidates on it that were running for a whole list of offices. Voters were permitted to leave a blank for any given office, but the candidates they chose all had to be of the same party. If a voter split the ticket and voted for candidates of different parties for different offices that automatically invalidated his vote.

The São Marcos Shavante thus faced a problem. They were staunchly pro-government. Was that not where their benefits came from? After all the Salesians too were tied in financially with the Shavante Project. But the opposition candidate for governor in their state of Mato Grosso was a priest and they had learned to be staunchly pro-priest. So Aniceto conducted classes, showing his people how to leave the vote for governor blank, so as not to vote against the priest, and then vote for all the other candidates on the government slate. It made no difference to them that their own Mario Juruna was an opposition candidate in Rio de Janeiro. Quite the contrary. They regarded Juruna as a renegade and a troublemaker and assured me that he would not be elected to anything at all, if he had to count on their votes. In the event Juruna was elected as the opposition swept all the major states, but the government avoided humiliation by retaining the governorships of an equal number of less populous, backwoods states including Mato Grosso where these Shavante cast their votes.

The actual voting was more dramatic then we had bargained for. On the eve of the election I was attending the men's council in the village when Aniceto formally demanded that I stand up because he wished to address me. I did so with mixed feelings. I knew I was about to be asked for something and nowadays Shavante did not request fish-hooks or ammunition for their rifles. They were more likely to demand a tape-recorder or a vehicle. Mario Juruna had even persuaded someone to provide a generator for his village. At least the São Marcos Shavante did not lack electricity, but it was quite conceivable that the request would be for something equally expensive. I had already noticed that Aniceto treated me like an outsider when he felt like it, but stressed my long-standing links, indeed my honorary kinship with the Shavante when it suited him. It was therefore quite a relief when I realized that he was only asking me to take a carload of Shavante into Barra do Garças to vote.

But I was bewildered. In this part of the world the only way to make sure you get your votes is to provide transport for your voters to get to town, and government pollsters had already promised to send a truck to take in all the Shavante that would vote for them and to feed them after they had cast their ballots. Why then was it necessary for us to go too? Because, Aniceto explained, he did not trust the government electioneers. Besides, if they came, they would get the Shavante into town late in the day. By that time the combination of election fever and rum would have turned the town into a place just spoiling for a fight. He did not want his people involved in any trouble. He wanted them to get in there, vote and get out again before the drinking and fighting set in. So he was commandeering other vehicles to do the job.

'All right,' I replied. 'We can set off just before dawn.'

'Negative,' answered Aniceto crisply. He was speaking Portuguese now to make sure there was no misunderstanding about the arrangements. 'You must be there at dawn, when the polls open.'

I was not thrilled at the prospect of driving over those difficult and unfamiliar trails in the dark and I particularly disliked the thought of the bridges over the numerous rivers we had to cross. They consisted of two logs, spaced more or less to coincide with the wheels of one's vehicle. One had to hit them cleanly and exactly, step on the accelerator and pray. Aniceto was quite firm however. There were plenty of young Shavante who knew the trails and they would even drive the Volkswagen for us if we wished. The only concession I wrung from him was that there were to be no more than nine people in the minibus besides the driver. I was not sure how much longer our vehicle would stand up to systematic overloading and Shavante men are hefty. Aniceto agreed and I went off to tell Pia what I had let myself in for.

Pia at once volunteered to drive, insisting that I should not waste my limited research time on a trip to Barra do Garças. I accepted her logic and her offer. She went out into the moonlight to get the bus ready and found it already occupied by fourteen Shavante who had somehow managed to squeeze into it. Nevertheless she set off cheerfully into the night. I could still see her tail lights when I remembered that, because of our detour through the sands on our way to the mission, she did not have enough gasoline to get her to Barra do Garças.

It turned out that one of the Shavante could indeed drive and he knew the road very well so he and Pia took turns at the wheel. They realized that their gasoline was low and hoped to fill up at a pump on the way to Barra, but it was predictably closed in the pre-dawn hour when they got there. Pia suggested that they wait until someone woke up and served them but the Shavante were impatient. They wanted to push on and take a chance. After all, one never knew exactly how much gas there was in the tank and the car might just make it into Barra before running dry. It did not. It hiccoughed and stopped twenty miles short of their destination. The Shavante were unperturbed. By now they had reached the main road and there was a fair amount of traffic going towards Barra on this election day. They all thumbed rides into town, promising Pia that they would get someone to come back to her with gasoline. Pia had little confidence in these assurances but had no alternative except to trust in them. So she made herself comfortable, as one learns to do in the backlands, and whiled away the time by writing up her notes. Sure enough her friend the driver came back two hours later with one of the electioneers and a can of gasoline.

In Barra do Garças she learned that Aniceto had had good grounds for his forebodings. Brazilians arriving early at one of the voting places had been outraged to find that it was already full of Indians lining up to cast their ballots. Some men had stalked out, saying that if they had to vote with a bunch of Indians, then they would not vote at all. The Shavante managed to vote without major incident, but now there was an impasse. Their patrons had promised them a feed after the vote but now, possibly fearing more disturbances if their Brazilian voters arrived to find scores of Indians pressing around the tables, they told the Shavante that they could not serve them till the afternoon. The Shavante were both indignant and hungry and Pia had no wish to wait till late afternoon before starting on the long journey back to the mission. So she invited all her passengers to eat in town at her expense so that they need not wait around any longer. Their Brazilian patron was mortified. 'Don't be like that, *senhora*,' he said, 'I am going to look after my Indian friends here'; but the hungry Shavante had already deserted him and were crowding into a local eating house.

We could see why the Indians needed their patrons. Towns like Barra do Garças were slowly becoming the focal points of their lives, yet without patrons trips to town were a miserable experience for them. They waited around, classically stolid, unwanted and with no place to go. Watching them, so out of place in the midst of their old ancestral lands, I was again reminded of the secular injustice from which they suffered. Some of them reacted boorishly, marching in where they were not wanted and demanding things instead of asking for them; but how else were they ever to get in, since they were so rarely invited? Yet we knew that Shavante guests could be very trying to non-Indians who did invite them. They are a proud people and we liked and admired them for the way they stood up to the world, but that did not make them any easier to live with. Anybody who has looked after a band of Shavante visitors needs saintly patience in order to put up with their demands.

They were poised between two worlds. They were urged on all sides to abandon their old, barbaric ways. Yet their own rich heritage still gave meaning to their lives. What was the alternative? They could become peons in Central Brazil and join the masses of the rural poor, aspiring to beer and rum in places like Matinha if they had money to spare and could get anyone to serve them. Or they could follow the missionaries. These at least offered them a vision of the future as vivid and compelling as the past they were told to abjure. Nobody mentioned the other option, that they might retain their lands and enter the Brazilian economy while modifying but not abandoning their own traditions. It was only a few bold Shavante who clung to that vision and they did not have many weapons in the battle for the souls of their fellow tribesmen.

We wondered how their distant cousins, the Sherente, were coping with the same problem. They had been struggling to maintain their own way of life against the encircling cattlemen for over a century. How were they faring now that Central Brazil had been 'opened up to development'? To try and answer that question Pia and I set out to revisit them two years after our return to the Shavante.

We drove up the Brasilia-Belem highway in the same old Volkswagen bus that we had used among the Shavante. The road had been the first step in the opening up of Central Brazil. Starting out from the new capital of Brasilia it goes due north through the backlands of the state of Goiás, cuts through the Amazonian

279

jungles in the state of Pará and ends in the city of Belem at the mouth of the Amazon itself. It was the first road linking the north of Brazil with the rest of the country and the first road to make access to Central Brazil relatively easy. Now it was paved for most of the way and lined for hundreds of miles with cattle ranches and little boom towns, with their blaring televisions and advertisements for farm equipment. The road runs roughly parallel to the Tocantins River. That traditional artery of Central Brazil is now a backwater, its infrequent motor boats mostly carrying cargo and not much of that, for why bother when it is so much easier and quicker to send it by truck? The towns on the east bank, away from the road, are withering now that they are approached overland with all the inconveniences of a ferry at the end of the journey. Even the towns on the west bank have to struggle to hold their own against new centers springing up along the road itself.

Tocantinia, being on the bank away from the road, was as sleepy as we had remembered it. We left it as quickly as we could and drove north to the Sherente. In the old days the river had so dominated our thinking and our mental map of the region that now it was hard to find our way over the tracery of overland trails. Besides, the old villages we knew had all moved. But the post was still there at the same place on the banks of the Tocantins and we headed for it to present our permits and begin our work. Even then we lost our way and were startled to find ourselves driving into a Sherente village whose huts were arranged in a perfect circle.

Traditional Sherente villages were laid out on a horseshoe plan, like those of the Shavante. Less traditional Sherente villages straggled without any particular plan, like neighboring Brazilian settlements. We had never seen a circular Sherente village before, but we had no time just then to reflect on their cosmic geometry. We were welcomed calmly, like returning relatives, by a man whom I mistook for Pedro who used to be chief of Porteiras. Surely Pedro would have aged more than this, I thought ruefully, before realizing that I was talking to Abel, Pedro's son. He explained to me that he was one of the two present chiefs of this village. They represented the two moieties of Sherente society and they occupied houses directly facing each other across the village circle.

I was pondering this unexpected traditionalism when I remembered that we had to go to the post. I explained this to Abel and assured him that we would return because I would like to spend some time in his village. Abel was startled by what he clearly felt would have been an abrupt departure. He suggested that we should stay a little longer, especially since his father was preparing to come over and greet us. At last Pedro appeared, spry as ever, only now with a full head of white hair. He was carrying a sword club and had dressed to greet us by opening the collar of his check shirt and putting on a cotton necklet with a hawk's feather sticking out at the nape. He leaned on the club and made us a formal speech of greeting.

I remembered the first time we had been formally greeted by a Sherente, by old Wakuke, now long dead. I remembered too how I had gone with so much trepidation to the village of Porteiras thirty years ago and how I had been warned against this very man, the bad chief of the most recalcitrant Sherente. I also remembered that I had come to like him and to sympathize with him as he fought to defend his people. Now, as he stood before me, old but unbowed, and pronounced the dramatic phrases and clicks of true Sherente oratory, I found myself deeply moved.

After the speech Pedro took off his necklet and slipped it onto me. Then he stood with Abel at his side and watched us as we drove away to the post.

We recognized the post by the tall familiar palm trees along the river bank, otherwise we would not have recognized it at all. Eduardo's modest little hut was gone. Instead we passed a neat house belonging to a Protestant missionary and then turned onto an approach that was lined with houses, some thatched, some built of brick and tile. There were vehicles and tractors parked by the main buildings. Modernization had clearly reached the Sherente too.

In fact they talked about little else. If the Shavante had been anxious about their Project, the Sherente were obsessed with theirs, indeed with a whole series of projects that they were constantly discussing, criticizing and trying to get in on. The old land disputes still festered. The leading citizens of Tocantinia still inveighed against the Indians as the source of all their woes, but the bigger ranchers had prudently accepted indemnities from the Indian service in order to leave the Sherente in uncontested possession of the northern part of the municipality. The smaller homesteaders predictably got nothing, and were simply ejected from what was now recognized as Sherente land. Meanwhile the Sherente who lived south of Tocantinia in a village called the

Funil, or funnel, were also left out of the provisional settlement. They claimed correctly that they had occupied their lands since time immemorial. In fact records going back two centuries show that there have always been Sherente there by the narrows of the Tocantins River. But the small population of the Funil had few resources now and even fewer allies. They were hemmed in by settlers who were pressing to have them evicted altogether, arguing that they should simply move to the Indian lands in the north. The Funil Sherente refused to move away from the bones of their ancestors. The Sherente to the north paid lip service to their plight, but gave them little support. They were too busy with their projects. These were a constant source of gossip and intrigue. The inhabitants of virtually every village claimed that other villages received more project funds from FUNAI than they did. To hear most people tell it, not only did the bulk of the project funds go elsewhere, but what little came to their village went to the chief and his friends and hardly benefitted them at all. The chiefs and their friends meanwhile complained that FUNAI was incredibly stingy with its funds and that they were constantly being dunned by villagers who refused to believe how little there was to go round.

In the past a Sherente faction that felt mistreated in its community would often secede and set up a new village. Now there was a new incentive to turn to this old solution—a new village could demand its own separate project from FUNAI. In the Village by the Post (now its official name) there was even a faction that wanted a separate project for itself without seceding. These were elders who considered the present chief Isaac a youthful upstart without much legitimacy and complained that none of the project funds reached them. Isaac retorted that the project funds were given for the modernization of Sherente agriculture and therefore were unlikely to be of direct benefit to men who had by now retired from farming anyway.

Isaac had launched his career through a happy accident many years ago. At that time he was employed by the Indian service at its post over on the Rio do Sono. The government that came to power in the years before the military coup passed legislation guaranteeing family benefits to those in its service. These were modest but welcome bonuses for people living in Brazilian cities. Isaac, who had never had a cash income and now received benefits for a wife and eight children, suddenly found himself well off by the standards of the backwoods. He therefore took care to keep in with the Indian service, and their support eventually propelled him into the chieftaincy of the most important Sherente village, the one with the most direct access to the benefits flowing from the government, the Village by the Post.

Meanwhile the elders cross-questioned me about the chances of a special project for them and about other sources of funds for the Sherente. One of them had just come back from a long reconnaissance to the big cities and was preparing to set out on his fund-raising travels again. He had heard that the World Bank funded projects and asked me for its address. He was disconcerted, but not deterred when I gave it to him, explaining that it was in Washington. He said he would find someone in Brasilia who would take his request to Washington. These canny old men seemed to think of the outside world as a vast treasure trove, where one could strike it rich if only one had the persistence to keep trying to find the right button to press. But there were plenty of people in the great cities of the world who had a similar attitude towards public funds, I reflected. What was remarkable about these Sherente elders was their determination and the crazy courage with which they faced the odds against them. Recently they had been lucky. Their people had been neither massacred nor dispossessed, though both seemed real possibilities at one time; and when they had made demands on FUNAI, the service had responded with projects.

But the idea of a project for the elders was more than an attempt to hit the jackpot on the part of some old men. It indicated the deep tensions between the older traditionalists and the younger generation.

'The elders,' one man explained to me, 'administer the past.'

It was a past that seemed increasingly irrelevant to younger Sherente, yet they were fascinated by it. When I read stories to them that had been dictated to me years previously by Suzaure, young men gathered round eagerly and asked if they could bring their cassettes and tape them.

'Nobody tells those old stories any more,' they explained.

It seemed that Pia's and my very presence, with our tales of the past and our books and pictures referring to a bygone era, was serving to kindle a miniature revival movement. But theirs was a fleeting interest. They enjoyed hearing about a time when the Sherente had been self-sufficient, when they had had a rich life of their own and had lived it according to their own lights. But they knew that those times were gone. Now they knew they lived at the edge of a larger world that decided their lives for better or worse and it was in that world that the real action was.

Or was it? I remembered the circular village that we had first arrived in and the traditional welcome that we had received. Pedro was of course one of the elders who 'administered the past,' but his son Abel was the chief. Abel must be about the same age as Isaac, yet he maintained Sherente tradition in a way that Isaac did not. Isaac's Village by the Post showed its modernity by being laid out in a straggling line of houses, rather than the circles and semi-circles of the old days. I asked about the circular village—it was not the only one among the modern Sherente—and learned that the Indians knew it did not correspond to the traditional lay-out hallowed by their ancestors. They had seen the circular villages of their neighbors the Kraho, whose customs were quite similar to those of the Sherente, and liked their uncompromising design. So, when the Sherente wanted to make a statement by building new villages in a traditional style, they built them in a circle. The circle was also the antithesis of the street. People in these parts spoke of 'going to the street' when they meant going into town. Town was a street of shops and bars. A circular village said something therefore about the Indian way as opposed to the Brazilian conception of the street.

The Sherente were quite clear about the contrast. Isaac explained that the Village by the Post was laid out like a street because that was more modern. Indeed there were Sherente in the circular villages who felt that they would have preferred a modern street instead, but they went along with the circle and much else that that implied.

The implications were not merely esthetic. Traditional Sherente villages were microcosms of their world. The moieties faced each other across them and the divisions of their society were daily synthesized in the central meetings of the men's council. All this was lost in the street. Modern villages did not provide the mechanisms for neutralizing opposition and creating harmony at the center. Their only focus was the chief's house, but meetings outside the chief's house were not the same. They were usually only attended by the chief's faction. Dissidents stayed at home to brood or gathered elsewhere to plot.

Nor were the differences only political, as we discovered when we stayed in Abel's village. It was the middle of the dry season, the traditional time for ceremonies among the Sherente. They were celebrating a name-giving. Pedro and our old friend Sizapi led the young men singing round the village at dawn and at dusk and sometimes during the night as well, as they prepared to bestow names on the girls.

Pia and I were only too happy to get up for this, since the nights were bitterly cold and the house that had been put at our disposal was an old one infested with cockroaches. They appeared as soon as the sun went down and swarmed over our dirt floor and all our possessions in a repulsive black mass. We rigged up makeshift plastic awnings over our hammocks so that they did not drop onto us out of the thatch, but they penetrated our defences and invaded our blankets at night. If we reached for a flashlight or a water bottle our hands would close over the wriggling creatures. By the early morning hours, cold and disgusted, we were only too happy to get up and go out to the fires in the middle of the village or, in my case, to join the singers.

I had provided a steer to be slaughtered for the ceremony, just as in the old days, only this time I could drive into Tocantinia and buy it and it was brought back to the village in the Indian post's own truck. The ceremonies climaxed in an enthusiastic log race, followed by a gargantuan feast of meat pies, huge hunks of beef wrapped in manioc and roasted in outdoor earth ovens.

I had also provided a steer for the Village by the Post, but they could not decide if or when they would perform their ceremony. In the mean time they used the steer for a proper 'civilized' party, as they called it. These parties are now the big occasions to which Sherente look forward. News of them travels far and Sherente come from distant villages to wherever the party is being held. For days nobody at the Post could think of anything else. The man who was officially giving the party was doing so in order to fulfill a vow made to one of the saints in the local Catholic tradition. His daughter, beautifully dressed and attended by maids of honor, was the party queen who had the privilege of carrying the banner of the saint from house to house and then planting it outside her father's house, where the party was to be held. There scores of men and women had been at work for days, clearing the ground, building a thatched extension to the house proper, erecting a huge pole for the saint's pennant, building earth ovens to cook the food, setting up trestle tables and at long last preparing the food and drink.

Few people waited to drink until the party started. Some had been drinking for days. Others started on the day of the party. An alcoholic haze hung over the village as people talked ominously, but with mounting excitement, of the fact that there was bound to be fighting at the party. There always was when people got drunk. The Protestant missionary prudently left, as he always did on these occasions, to sit out the binge in his house in To-cantinia. All day people came streaming in from other villages. They came on foot, on horseback and in the vehicles of the Indian service. Jacinto, the first Sherente chief whom we had ever met, was brought in by cart. He had long ago taken to the bottle in earnest and now lay around, a semi-invalid whose family hauled him to whichever party they could reach.

Food was served as the sun began to set. Groups were called to the tables, village by village, and miraculously there seemed to be no shortage of things to eat. Plenty of Brazilians had also come to the party, dressed in their best, and they were buying food and drink at the tables that sold them. Night fell and musicians struck up on drums and accordion when Pedro, the old chief, wearing an authoritative white sombrero as well as his eagle's feather, stepped onto the dance floor and called for silence. Remarkably, he got it. He made a speech in the Sherente language, admonishing everyone to enjoy themselves, but not to get drunk and above all not to fight. He then repeated his speech in Portuguese and with that the party formally got under way.

Pedro had to repeat his speech in the early hours of the morning. He spoke with greater vehemence this time for people were weaving drunkenly about on the dance floor and his own oratory was helped by the rum that he himself had knocked back.

I stood at the edge of the affair among the wallflowers and the old men. The latter constantly asked me whether I liked the party. I soon learned that this was not a social question. They wanted to know if I approved. I tried to be non-committal. Most of them then told me in no uncertain terms that they did not approve, especially of this wiggling, two-by-two dancing that only led to fighting and trouble. They preferred the traditional group dancing of their own people.

In the middle of the night I was cornered by Jacinto's wife, in whose household Pia and I had gone hungry. She told me that Jacinto was asking for me and I went in to see the old man, who was lying in the hut of one of his children. It was immediately clear that it was she rather than he who wanted to see me. Jacinto lay there, mumbling sheepishly, while his wife embarrassed us both.

'Look at him,' she said. 'He is really in a sad state. He is pathetic. Anybody else would have left him, but not me. I am a good wife. I have pity on him. I look after him. But we have nothing. We need clothes and food, coffee and salt.'

I was squirming with embarrassment now so she pressed home her advantage.

'Have pity on him. Give him something. You have so much. Give us some clothes. We need blankets. . . .'

I gave her some money just so that I could get away, knowing full well that it would be spent on rum before the night was out. In fact there were not many fights that night. The only man with a deep knife wound was a Brazilian who was hurt well away from the party. Next morning young men with crashing hangovers wandered about the village saying what a good party it had been, for there had hardly been any fighting. They sounded a bit disappointed.

I felt sorry for the younger generation. Sherente culture, that intricate creation that had served their people since the beginning of time, was now little more than an ethnic marker for them, something they considered of folkloric interest. It did serve to distinguish them from the surrounding Brazilians, but they were no longer quite sure whether they wanted to be distinguished in that way. Yet the resources of their own tradition contrasted sharply with their present intellectual and spiritual impoverishment. Their culture had been developed over the centuries by the contributions of speculative individuals. Where were the modern descendants of those heroes? I suspected that they still existed among the Sherente I knew, but for them the times were out of joint. Their intellect and imagination could hardly flower in a world where Sherente ways were despised as barbarous and foolish by outsiders powerful enough to make the Sherente themselves have doubts. So Sherente innovators are stifled and Sherente culture begins to wither. But what can the rising generation put in its place? They seem to be agnostics, scorning the old ways and mistrusting the new, hoping only for women, football and jobs in the Indian service.

It was time for us to leave. We were now constantly being asked when we would return and I was more than usually troubled by the question. I suspected that I might not have the chance to come back again, that it might never again 'make sense' for me to do so. So I answered awkwardly, as if I were breaking my faith with the Sherente, though I had never promised to come back, and suspected that most of them would not care very much one way or the other. Then why was our return of any consequence to them? Perhaps it was because they knew of our interest in them. In that sense they knew that we like them and such people are so rare that they are remembered. They wanted us back because we were friends from an outside world that is so alien and, on the whole, so hostile to them.

As usual we gave away all of our equipment when we left. Our blankets were the last and most valuable gifts that we had to bestow. They had been promised to the elders, and Sizapi and Jovino were waiting to receive them in the dawn hour of the day we were to leave. They saw us impassively into the old bus and then bade us wait a moment.

They wanted to sing for us. The classic Sherente songs rose plaintively into the cold morning air, as the old men stood and chanted with our blankets wrapped around them. Then we sadly drowned them out with the noise of our motor.

Index